activity-based statistics

SECOND EDITION

STUDENT GUIDE

Richard L. Scheaffer, University of Florida
Ann Watkins, California State University, Northridge
Jeffrey Witmer, Oberlin College
Mrudulla Gnanadesikan, Retired, Fairleigh Dickinson University

Second Edition Revised by
Tim Erickson, eeps media

JOHN WILEY & SONS, INC.
New York / Chichester / Weinheim / Brisbane / Singapore / Toronto

Published by John Wiley & Sons, Hoboken, NJ

For general information on our other products and services, please contact our Customer Care Department within the United States at 800-762-2974, outside the United States at 317-572-3993 or fax 317-572-4002.

For more information about Wiley products, visit our website at www.wiley.com.

ISBN: 978-0-470-41209-1

Development Editor: Allyndreth Cassidy
Production Director: Diana Jean Ray
Production Project Manager: Michele Julien
Project Management, Composition, Technical Art: Graphic World Inc.
Text Designer: Amy Feldman
Art and Design Coordinator: Kavitha Becker
Cover Designer: John Nedwidek
Cover Photo Credit: Paul Wenham-Clark/Pictor International/PictureQuest

Executive Editor: Richard Bonacci
General Manager: Mike Simpson
Publisher: Steven Rasmussen

Original *Activity-Based Statistics Project* advisory committee

Donald Bentley, Pomona College

Gail Burrill, Greenfield, WI

George Cobb, Mt. Holyoke College

Elizabeth Eltinge, Texas A & M University

Joan Garfield, University of Minnesota

James Landwehr, AT&T Bell Laboratories

Donald Paxson, Florida Power and Light

Dennis Pearl, Ohio State University

Judith Singer, Harvard University

Frank Soler, De Anza College

W. Robert Stephenson, Iowa State University

Evaluator

Mary E. Huba, Iowa State University

CONTENTS

III. ANTICIPATING PATTERNS

IV. STATISTICAL INFERENCE

V. PROJECTS

PREFACE

Since the publication of the first edition of *Activity-Based Statistics* by Springer-Verlag, New York, in 1996, interest in an activity-based approach to teaching introductory and AP statistics has grown considerably and has become part of the established pedagogy for teaching these courses. The intent of the original grant from the National Science Foundation for the Activity-Based Statistics Project (ABSP) was to provide instructors of introductory statistics with a great set of basic, hands-on activities that could be comfortably integrated into any introductory statistics course. Because these activities were designed to be used without the aid of any technology tools, their implementation was made easy for any instructor without regard to budget concerns or technology proficiency. By making the activities fun and engaging, the authors are promoting the idea that statistics is, indeed, an experiential science.

This second edition of *Activity-Based Statistics,* revised by Tim Erickson of eeps media, remains true to the vision of the original, highly respected author team: a set of basic, hands-on activities that can be completed without the use of technology. Furthermore, these activities are interesting and conceptually rich and cover the AP statistics and college introductory course curricula.

Incorporating feedback from the numerous users of the first edition and the reviewers of the drafts of the second edition, the authors for this revision:

- Updated the examples and data
- Added two totally new activities, *V Is for Variation: How Far Are You from the Mean?* and *The Lazy Student,* an activity that explores how spread changes when you add random variables
- Improved *Instructor Resources* by embedding the instructor notes within the student material and clearly distinguishing between the two
- Introduced optional technology extensions that allow instructors to replicate, improve, or extend the basic activity using one of several technology tools
- Created a sampling of activity extensions in Minitab®, Excel®, and Fathom Dynamic Statistics™ with accompanying instructor demonstrations and data sets

TO THE STUDENT

Activity-Based Statistics was written to help you enhance your learning of statistics. You are most likely using this text as a supplement to your core statistics textbook. The idea of incorporating *Activity-Based Statistics* into your course is to show you statistics in context. Once you have mastered statistical concepts, it is a good idea to work through the activities so that you can see statistics outside of your classroom.

Activity-Based Statistics includes more than 45 activities that can be used in the classroom on a day-to-day basis and five projects at the end of the book that your instructor might have you complete over a long period. Each activity is structured similarly. You will always find a short scenario that introduces the statistical concepts in a real-world setting. Next, you will note the "Objectives" of the activity, a "Question" about the scenario, and the "Prerequisites." Activities end with "Assessment Questions" that reinforce your understanding, a "Wrap-Up" that summarizes the activity, and "Extensions" for continuing with the themes presented in the activity.

Throughout this text you will be gathering data, analyzing data, making observations, and drawing conclusions. You will often be asked to work collaboratively. Occasionally your instructor will place you in a group, and it is important that you follow directions carefully and participate fully. You will often have to read the activities closely, and you will be required to produce written responses to some of the questions.

Your instructor has chosen to incorporate *Activity-Based Statistics*, second edition, into your course because he or she believes in building real statistical understanding while moving away from the myth that statistics is dull and complex. By the end of this course, you too will hopefully believe that statistics can be fun and is applicable outside of the classroom.

First Edition Acknowledgments

The persons listed as authors on this work are greatly indebted to many other teachers of statistics who have willingly shared their successes and failures in the use of activities in the classroom. In particular, the Advisory Committee, listed inside the cover page, contributed their own ideas for activities as well as countless hours for review and class-testing of many of the activities that finally were selected for this publication. Special thanks are directed to George Cobb, who invented and wrote the activity on Gummy Bears in Space, which has turned out to be one of the most popular in the set. Mary Huba provided valuable information from users, both teachers and students, of the activities while they were in the developmental stage, and her work led to great improvements. Joan Garfield is due special thanks for writing the Assessment section of the *Instructor Resources.* It is not often that a project receives so many valuable contributions from such a variety of interested parties; we thank you, one and all.

Second Edition Acknowledgments

We would like to thank the reviewers who took time out of their busy schedules to review the revised material for the second edition and who provided excellent commentary on the changes that were made.

Amanda Bertagnolli-Comstock, Bishop State Community College

Jon Cryer, University of Iowa

Anne Jowsey, Niagara County Community College

Steve Patch, University of North Carolina-Asheville

Paul J. Roback, Connecticut College

Cathy Zucco-Teveloff, Trinity College

I

Exploring Data

A Living Histogram

How to build a histogram out of people.

Someone once said that statistics was the art of making decisions in the face of uncertainty. Before we attempt to meet that lofty goal, however, we must figure out how to look at the data we will use to make those decisions. As you will find in this course, we use data that *vary:* The numbers are not all the same. Variation is at the heart of statistics, and it's what makes those decisions-in-the-face-of-uncertainty so difficult.

How do we represent variation? You may be used to graphs in mathematics classes that display unchanging, solid relationships. But here, in statistics, we have to show how the numbers are different; we have to show the *distribution* of the numbers. The distribution tells us not only what numbers we're looking at but where the values are bunched up and where they are rare. A *histogram* is a special kind of graph we can use to show a distribution.

Objectives

This activity gives you a vivid reference for how to make a histogram. It also reinforces general ideas about distributions: shape, center, and spread.

Prerequisites

None.

Activity

In this activity, you and your class are going to construct a histogram of your hand spans. (You could use other quantities; hand span is just one of many possibilities.)

1. Work with your colleagues to measure the maximum distance from your thumb to your little finger, with your hand stretched out. Each student should find this distance to the nearest tenth of a centimeter. (A 25-cm ruler is provided at the back of the book.)

0 cm 1 2 3 4 5 6 7 8 9 10 11 12 13 14 15 16 17 18

Figure 1

2. Through discussion, find the largest and smallest values in the class.
3. Your instructor will move you to a large open area.
4. The instructor will make labels. Line up single file behind the label that includes your own hand span.
5. Look at the human histogram you have just made to see its features.

Assessment Questions

These questions are based on the following histogram of the ages of 500 people from northwestern Vermont (a region that includes Burlington). These data are from the 1990 U.S. Census.

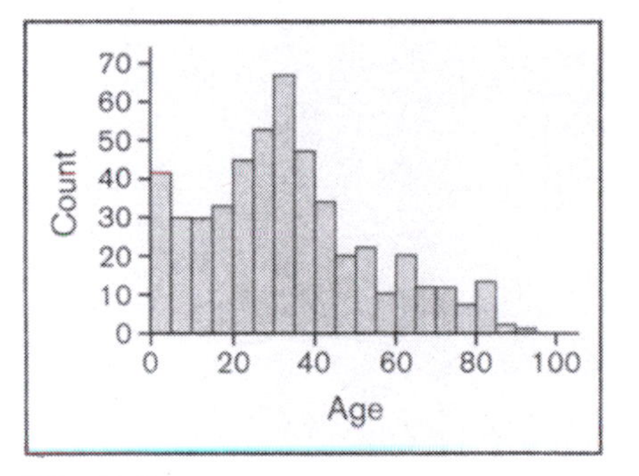

Figure 2

1. Describe the distribution you see. Explain any features you can.
2. The tallest bin is of people between the ages of 30 and 35. When were these people born?
3. This histogram has bins that are 5 years wide. Resketch this histogram with a bin size of 10 years.
4. Sketch how you think the distribution looks in the year 2000 Census.

Getting to Know the Class

Surveying the class. Exploratory data analysis. Displaying categorical and quantitative data.

No two people are alike. Some people are taller than others. Some people love classical music, and others do not. Indeed, people vary from one to another in many interesting ways. A good way to begin thinking about variability is to collect and then analyze data from those around you. In this activity we will collect data from members of the class. This will lead to some graphical methods of data analysis and a discussion of topics such as the wording of questions. After you have compiled the data, you can explore them and see where you fit within the class distribution of the variables measured.

Question

How do you compare with the other students in the class?

Objectives

This activity teaches some of the tools of exploratory data analysis and addresses some concepts in measurement.

Prerequisites

None.

Activity

Your instructor will give you a questionnaire (page 7 is a sample) to answer. Please answer the questions anonymously; feel free to skip any question you feel uncomfortable answering. Your class will collect these data for educational purposes. They will form a database for the class to analyze.

1. Complete the questionnaire. Turn it in when you are finished.
2. Your instructor will read the responses to one question from the questionnaires. Write the data down.
3. Create a graphical display of the data, working together in small groups or pairs.
4. Share your displays.
5. Discuss these questions:
   - **a.** Which displays use all of the data?
   - **b.** What information is lost in each display?
   - **c.** Which displays show you the distribution?
   - **d.** Which involve or let you calculate a measure of center?
   - **e.** Which is the easiest to make?
   - **f.** Which is the hardest to make?
6. Now pick a categorical variable: sex. Quickly graph this variable, and share the displays when you're done.
7. Discuss:
   - **a.** What's fundamentally different about this (categorical) variable compared with the previous (quantitative) variable?
   - **b.** Which displays are similar?
   - **c.** Which displays do not work with a categorical variable? Which do not work with a quantitative variable?
   - **d.** Why?
8. Discuss: Suppose we had a graph of pulse rate and a graph of sex and we had a conjecture that women generally had a slower pulse rate than men. Could we use those graphs to investigate that conjecture?
9. Work with your group to come up with a conjecture about the data—how two variables might be related to one another—and develop a graphical display to support or refute that conjecture. This should be "presentable," that is, on a transparency, on a poster, or in a computer file for projection.
10. Share your conjectures and displays with the class; discuss the challenges of making and interpreting the graphs.

Wrap-Up

1. Discuss what you learned about exploratory data analysis during this activity.
2. Discuss what you learned about measurement during this activity.
3. Suppose you wanted to know what proportion of the students at your college are vegetarians. How would you construct a question that would give you the information you need? What kinds of things could go wrong in writing the question? Would your question provide a valid and reliable measure of whether the student is a vegetarian? Would your results tell you about anything else, such as students' political leanings or health awareness?

Assessment Questions

1. Someone says that smokers tend to have a higher pulse rate. Using the class data, make a graphical display to illuminate that conjecture. What conclusions do you draw?
2. Someone says that people who carry a lot of change are likely to have more CDs. Using the class data, make a graphical display to illuminate that conjecture. What conclusions do you draw?
3. Suppose you have two categorical variables. What kind of display can you use to help investigate relationships between them? Give an example of two categorical variables that you think are related. How do you think they are related? Devise a plausible display that shows that relationship.

Questionnaire

1. What is your height *in inches*? ______
2. What is your pulse rate in beats per minute? ______
3. Are you male or female? ______
4. How much did you spend on your last haircut (including any tip)? ______
5. How much did you spend on your last restaurant meal (including any tip)? ______
6. How much money do you have with you right now, in change only? (Report the combined value of your change, not the number of coins.) ______
7. Choose a random number in the range 1 to 20. ______
8. Do you have a job in which you work at least 10 hours per week? ______
9. How many CDs do you own? ______
10. Do you smoke? ______
11. What time did you go to bed last night? ______
12. How old do you think the professor is? ______
13. What percentage of students at this school know the capital of Belarus? ______
14. How did you get to class/school today (for example, via foot, bike, car)? ______

A Living Box Plot

How to construct a box-and-whisker plot with people. Shape, center, and spread.

Much of the mathematics we study was invented more than 300 years ago. But that's not the case with statistics. In the 1970s, an American statistician named John Tukey (who coined the word *bit* for "binary digit") invented the box and whisker plot. Also simply called a *box plot,* this display helps us describe, visualize, and compare distributions.

If you were born after 1980, you are in the first generation of students ever to learn about Tukey's ingenious displays. In addition to the box plot, you have probably used (or will soon use) his stem and leaf diagram, which is especially well suited for making a quick plot of a distribution when you have to do it by hand.

Objectives

This activity reinforces the process of constructing a box plot and provides a vivid exercise to help students remember it. And as with *A Living Histogram* (page 2), it also gives you a chance to reinforce general ideas about distributions: shape, center, and spread.

Prerequisites

You should know how to find quartiles and how to construct a box plot using paper and pencil. Note that you can do this activity in the same class session in which you first learn about box plots.

Activity

1. In this activity, your class is going to construct a box plot of the times each student went to sleep last night. (You could use other data; bedtime is one of many possibilities.)
2. As a group, find out who went to sleep the earliest.
3. Find out who went to sleep the latest.
4. Place evenly spaced marks on the floor to represent the hours that encompass the earliest and latest times. That is, make a continuous axis.
5. Position yourselves at the appropriate points along the time line, according to when you went to sleep the previous night. If two students went to sleep at the same time, then one should stand behind the other.
6. Have the median person step forward.
7. Have the first quartile step forward.
8. Have the third quartile step forward.
9. Compute the interquartile range.
10. Outliers should step forward and turn sideways so that they look different from the nonoutliers.

Assessment Questions

Following are the violent crime rates, as of 1999, of 23 of the largest cities in the United States (in incidents per year per 100,000 population; data from the FBI's Uniform Crime Reporting System, found at http://www.fbi.gov/ucr/99cius.htm):

City	Violent Crime Rate	City	Violent Crime Rate
Austin	529	Milwaukee	1043
Boston	1302	Minneapolis–St. Paul	1161
Columbus	855	Nashville	1607
Dallas	1414	New York	1063
Detroit	2254	Philadelphia	1604
El Paso	686	Phoenix	832
Honolulu County	254	San Antonio	561
Houston	1187	San Diego	598
Indianapolis	1016	San Francisco	866
Jacksonville	1034	San Jose	581
Las Vegas	665	Seattle	767
Los Angeles	1283		

Table 1

1. Construct a box plot of these data.
2. Describe the distribution shown by the box plot in step 1.
3. Based on these data, how large or how small would a crime rate have to be to be an outlier?
4. Portland, Oregon, had a rate of 1236. Add this city to the 23 cities listed in Table 1 and construct a box plot of the 24 data values.

V Is for Variation: How Far Are You from the Mean?

Measuring variability in data.

Think about the students in your classroom. Now imagine a family get-together. It's possible that these two groups would have the same number of people. It's even possible that they would have the same average age. But chances are, the ages in the family gathering vary more than those in the classroom—they're more spread out. In this activity, you'll take measurements of your hands and use them to compare groups. Some groups will have bigger hands, some smaller. But the issue here is, some groups will have a bigger *spread*. How much bigger? We have to figure out a way to determine that.

Question

How can we describe the variability of data *quantitatively*?

Objectives

In this activity, you will learn how to figure out a way to measure variability in data and learn how that measure of variability reflects the actual data sets.

Prerequisites

You should know how to find the mean and median of a set of numbers and how to make a dot plot.

Activity

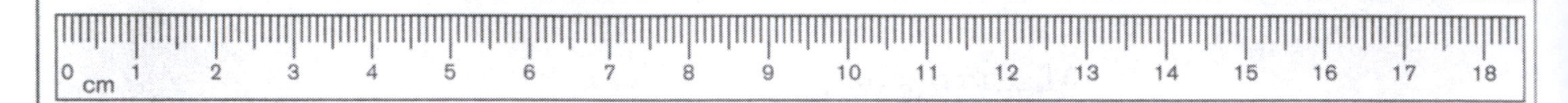

Figure 1

1. In groups, each student measures his or her "V-span" as follows: With the palm of your "writing" hand on a flat surface, make a "V" between the index and middle fingers. Measure the distance from the outside of your index finger to the outside of your middle finger when you spread them as far as possible. Measure to the nearest tenth of a centimeter. Use either the ruler above or the ruler at the back of the book.
2. Compute your group's median V-span.
3. Make a dot plot of your measurements. You should write names or initials above the dots to identify the cases and mark the median with a wedge △ below the number line.

You will now develop a way to quantify how spread out the measurements are by generating a *statistic*—a single number—to describe the spread of the measurements in each group.

4. Calculate the absolute value of the difference between your V-span and the median of your group.
5. Make a second plot, this time a dot plot of these differences from the median. Again, label the dots with names or initials.
6. Using the idea of differences from the median, calculate a number that gives a "typical" distance from the median. Indicate it graphically on both of your plots.

Wrap-Up

1. Imagine again the class and the imaginary family gathering. Estimate the median age of the two groups and the mean absolute difference (MAD)—in age—of the two groups.
2. Thinking back to the class and the hand measurements, what sorts of groups tended to have small MADs? What sorts of groups had large MADs? Can you explain these relationships?
3. Write why it might be important to have a quantitative measure of spread.

Extensions

1. Using a small data set, compute the MDFM—the mean difference from the *mean* without taking the absolute value first. What do you get? What could you use this measure for?

2. Learn about the standard deviation and discuss how it is different from the MAD.

3. We could calculate the median V-span of the entire class, calculate everyone's difference from that median, and thereby figure out the MAD of the V-spans for the whole class. But suppose we lost the individual V-spans and had only the MADs for the *groups*. In that case, could we still calculate that whole-class MAD? If so, how would we do that? If not, why not? Also if not, what could we tell about the whole-class MAD from the group MADs?

Assessment Questions

1. Calculate the MAD of these numbers:
   $\{1, 2, 3, 3, 3, 4, 6, 8, 10, 14, 20\}$

2. Mavis likes this measure of spread: Just take the maximum value and subtract the minimum value. The result represents the entire range and shows how spread out the data are. To get a typical distance, divide that by 2. What are the advantages and disadvantages of Mavis's method?

3. Invent two sets of numbers that have the same median and the same range but different values for the MAD. Describe what that difference in the MADs tells you about differences between the two sets of numbers.

Matching Plots to Variables

Connecting our knowledge of real-life distributions to their graphs.

Some families have very large incomes relative to the rest of the population. But these families are few in number, and the larger the income is, the fewer the families who earn it. Thus, if we were to collect family income data from a sample of Americans and then construct a histogram, we would expect the histogram to be skewed toward larger values. That is, by thinking about the variable, we can imagine the shape of the histogram without actually collecting the data. Can we do the same with other variables?

Question

Can we deduce the likely shapes of the histograms of each of several variables?

Objectives

In this activity, you will learn how features of distributions are manifested in graphs of the data. After completing this activity, you should be able to sketch the shape of the histogram for a variable by thinking about the nature of the data.

Prerequisites

You should be familiar with box plots and histograms.

Activity

1. Warm-up

   Consider the following two variables:

   a. Age at death of a sample of 34 persons

   b. The last digit in the social security number of each of 40 students

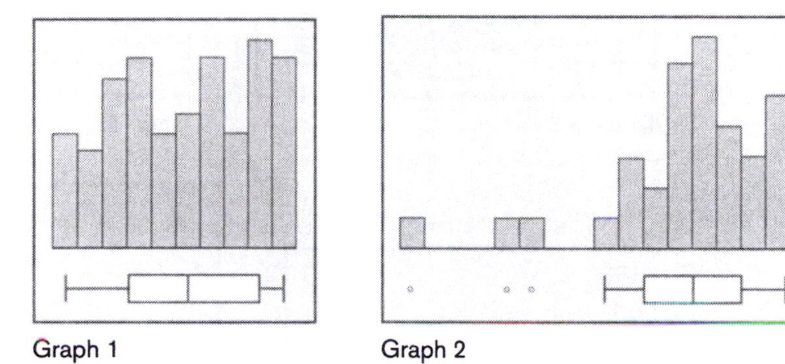

Graph 1 Graph 2

   We wish to match these variables to their graphs.

   We know that there are relatively few deaths among young people; the death rate increases with age. Thus, we would expect the histogram of the age-at-death data to be skewed to the left, with most of the observations being in a large group toward the right of the graph. We expect the histogram to have a small left-hand tail. Hence, we match a with 2. On the other hand, the social security data should have a distribution that is close to uniform on the integers 0, 1, . . . , 9. Thus, we match b with 1.

2. Main activity

   Consider the following list of variables and graphs:

   a. Scores on a fairly easy examination in statistics

   b. Number of menstrual cycles required to achieve pregnancy for a sample of women who attempted to get pregnant (Note that these data were self-reported from memory. Data from S. Harlap and H. Baras [1984], "Conception—waits in fertile women after stopping oral contraceptives," *Int. J. Fertility*, 29:73–80.)

   c. Heights of a group of college students

   d. Numbers of medals won by medal-winning countries in the 2000 Sydney Olympics

   e. SAT scores for a group of college students

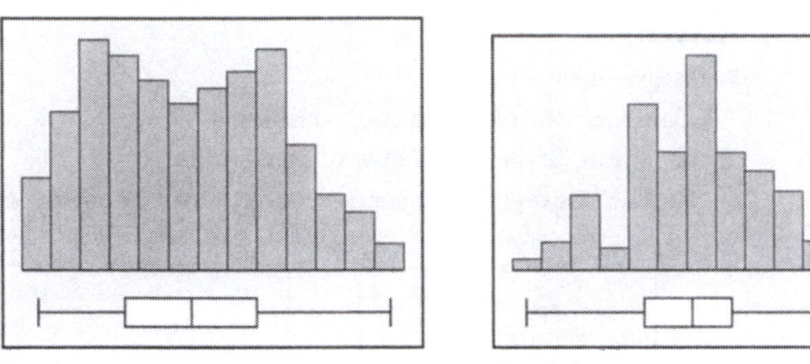

Graph 1

Graph 2

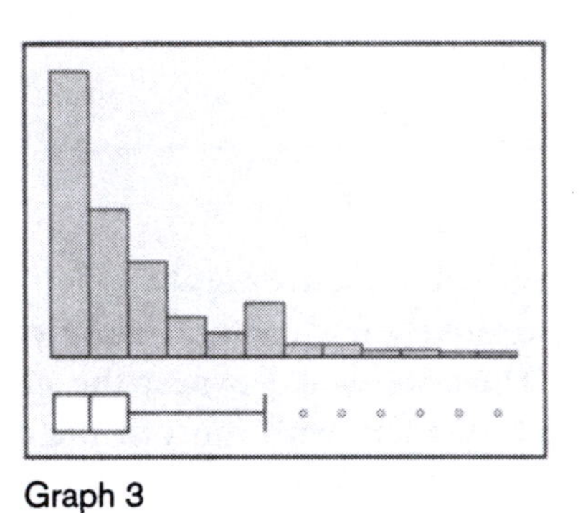

Graph 3

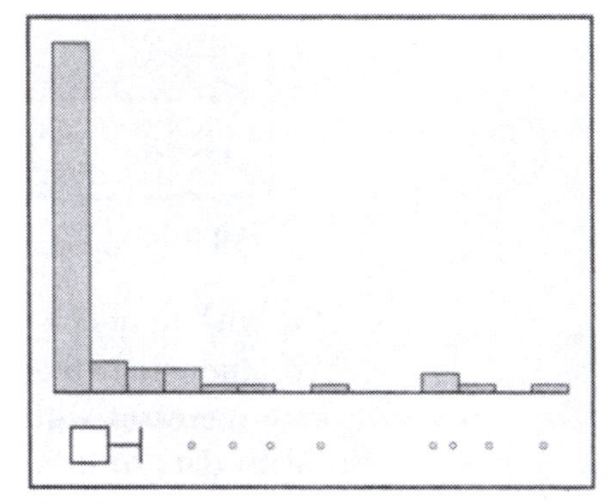

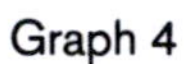

Graph 4

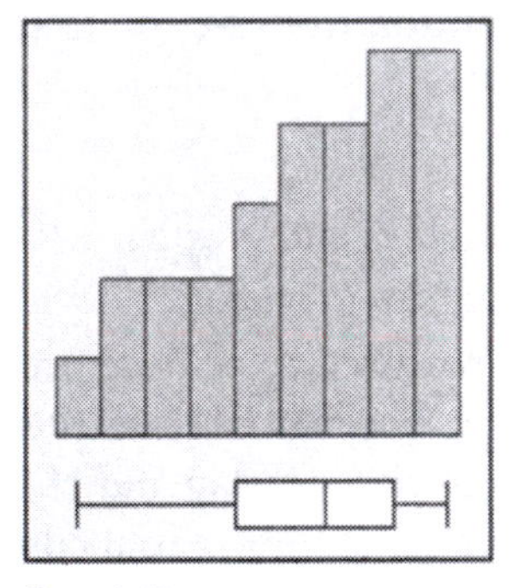

Graph 5

Use your knowledge of the variables (that is, ask yourself what the distribution should look like) and the knowledge in your group to match the variables with the graphs.

Wrap-Up

1. Write a brief summary of what you learned in this activity about how features of distributions are related to graphs of the data.
2. Name two variables that have symmetric distributions, two that have distributions skewed to the right (toward higher values), and two that have distributions skewed to the left (toward lower values).

Extensions

1. After you have matched the graphs to the variables, describe each distribution and any unusual features. For example, is the distribution skewed to the right? Is it symmetric? Are there outliers? Why might this be so?
2. In each case, estimate whether the mean is greater than, less than, or equal to the median. Explain your reasoning.

Assessment Questions

1. Consider the selling prices of houses sold in the United States during the past year. Sketch a histogram for this variable.
2. Following is a histogram for some data. Name a variable that might have led to this histogram. Explain why the histogram for your variable would have this shape.

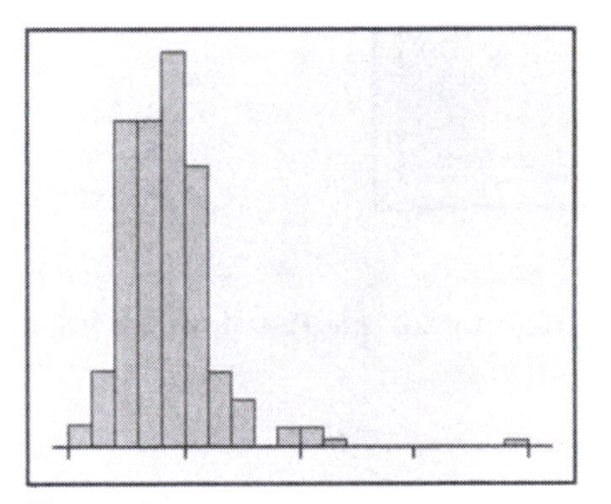

Figure 1

3. Repeat question 2, but this time name a variable that not only has the right shape but also is measured in the right units. That is, name a variable for which the smallest observation is near zero, the largest observation is between 75 and 100, and the shape is as indicated. Explain your reasoning.

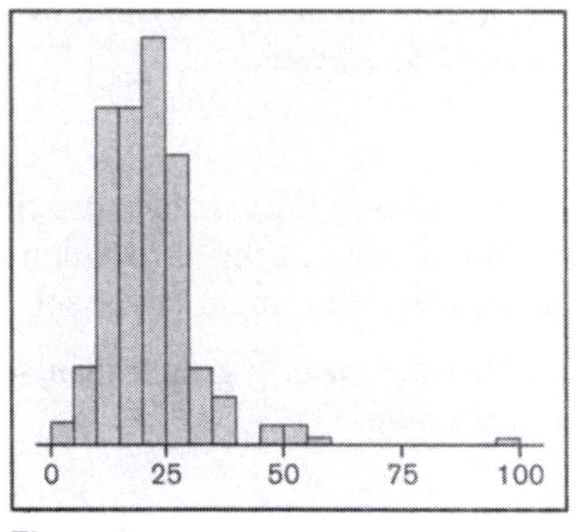

Figure 2

4. Repeat question 2 for the following histogram.

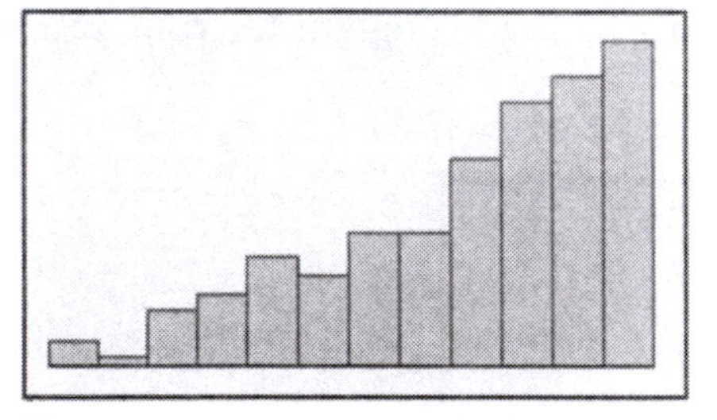

Figure 3

5. Looking again at the histogram for question 4, which is larger, the mean or the median? Explain how you know.

Matching Statistics to Plots

Matching summary statistics to graphs of distributions. How the mean can differ from the median.

You may have heard that the average lawyer earns a lot of money; at the same time, the *median* income of lawyers is more modest. Or consider professional sports. One year, the average team in the National Football League made a sizable profit while the median team barely broke even. People often confuse the mean and the median of a distribution. How are these statistics related to the shape of the distribution?

Question

Can we estimate the mean, median, and standard deviation of a distribution by looking at the histogram?

Objectives

This activity explains how summary statistics are related to graphs of data and how box plots are related to histograms. After completing this activity, you should be able to recognize when and how the mean of a distribution differs from the median. You should also be able to sketch an accurate box plot of a distribution after seeing the histogram and vice versa.

Prerequisites

You should be familiar with box plots and histograms, as well as with the concepts of mean, median, and standard deviation. You need not know how to compute these values.

Activity

1. Consider the following group of histograms and summary statistics. Each of the variables (1–6) corresponds to one of the histograms.

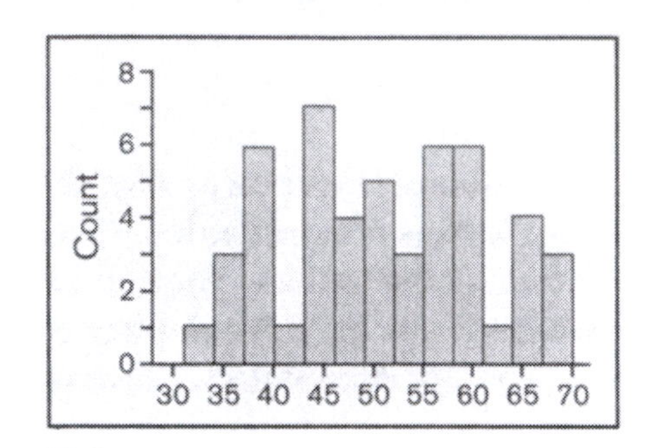

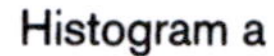
Histogram a

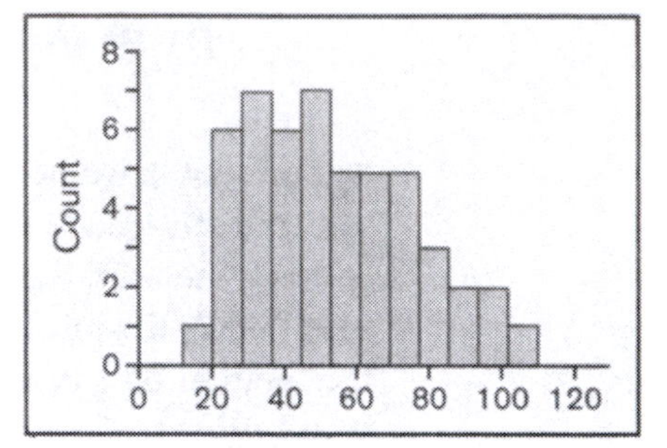

Histogram b

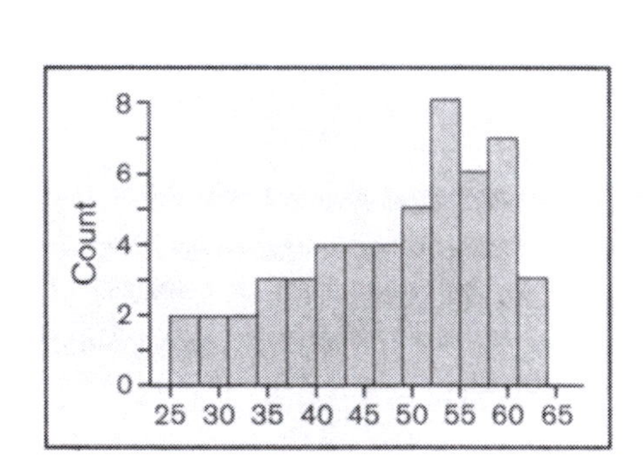

Histogram c

Histogram d

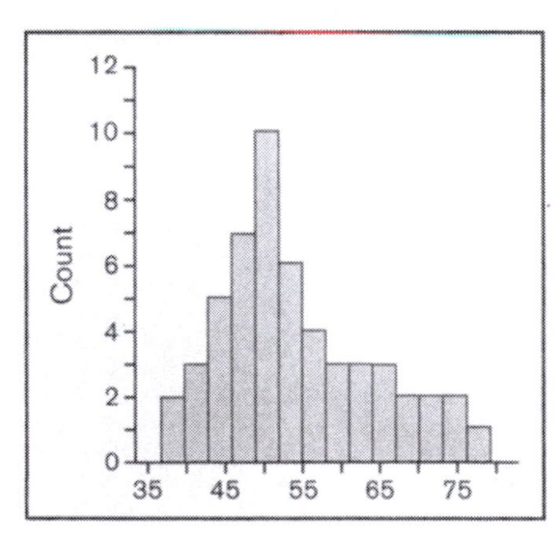

Histogram e

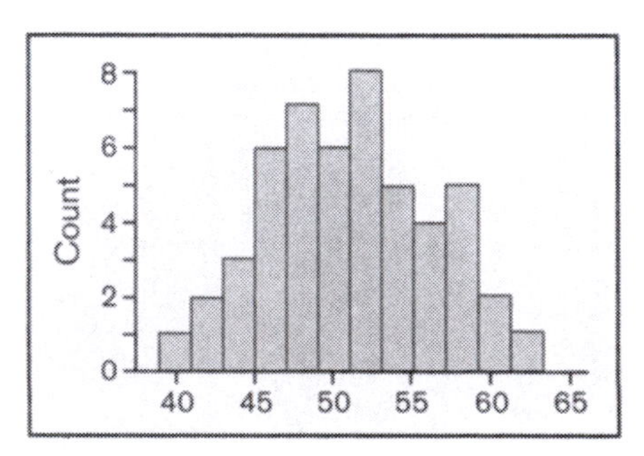

Histogram f

Write the letter of the histogram next to the appropriate variable number in Table 1 and explain how you made your choices.

Variable	Mean	Median	Standard Deviation
1	60	50	10
2	50	50	15
3	53	50	10
4	53	50	20
5	47	50	10
6	50	50	5

Table 1

2. Consider the following group of histograms and box plots.

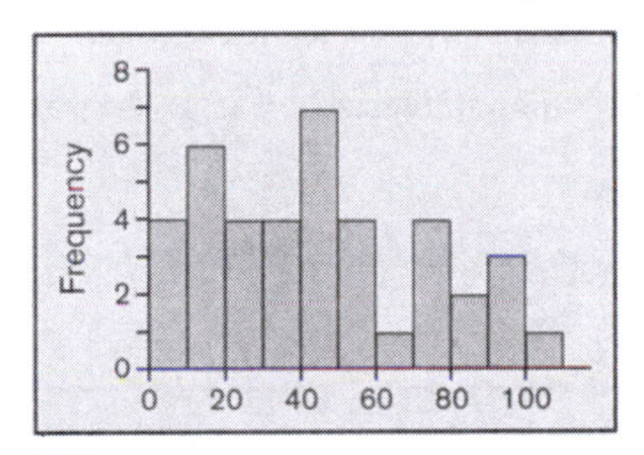

Histogram a

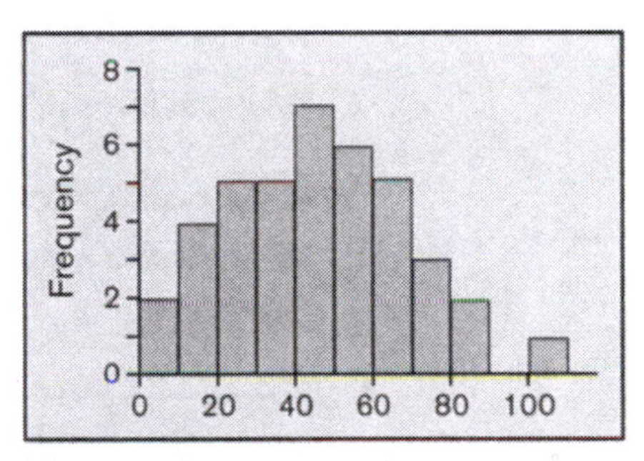

Histogram b

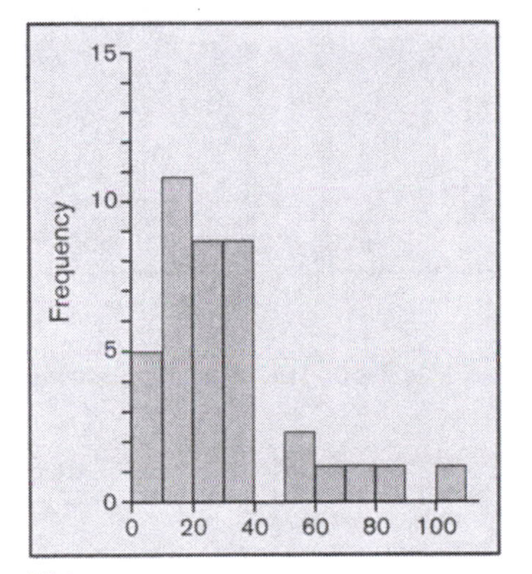

Histogram c

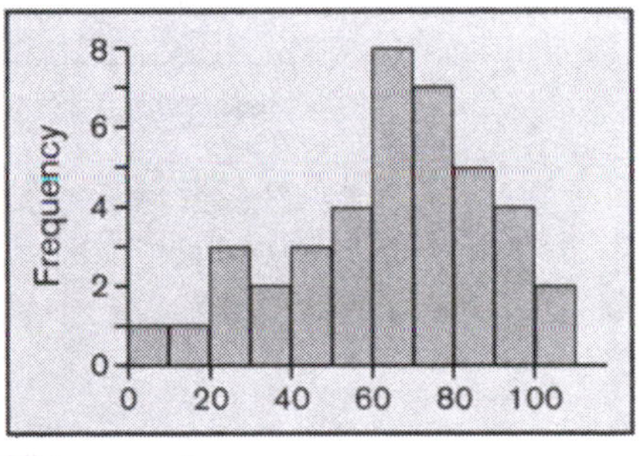

Histogram d

Box plot 1

Box plot 2

Box plot 3

Box plot 4

Each box plot corresponds to one of the histograms. Match the box plots to the histograms and explain how you made your choices.

Wrap-Up

1. What features of a distribution determine whether the mean and the median will be similar? When does the mean exceed the median?
2. What features of a distribution influence how large the standard deviation is?

Assessment Questions

1. Estimate the mean, median, and standard deviation of each of the distributions graphed here.

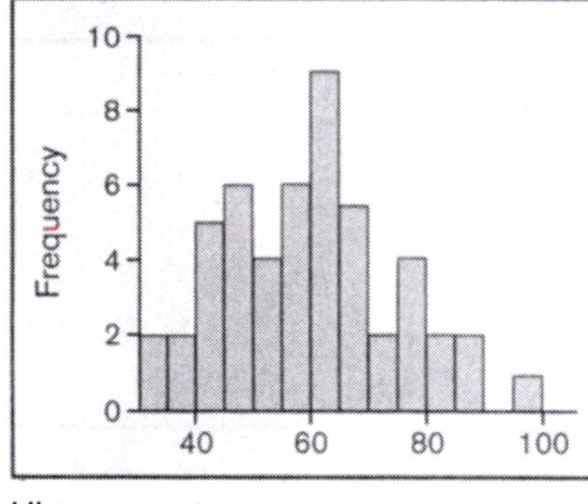

Histogram a

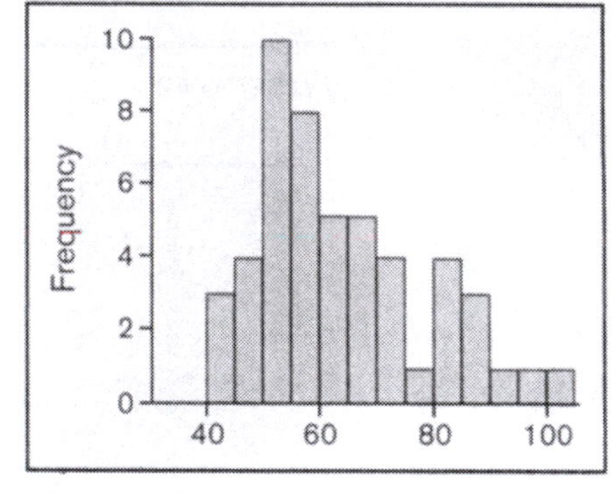

Histogram b

2. Sketch the histogram of a variable for which the mean is greater than the median.

Variation in Measurement

Collecting measurement data and looking at its distribution.

Wouldn't it be dull to look out your window and see people who were all the same size, cars that were all the same color, or trees that were all the same shape? Virtually everything around us is subject to variability. We tend to like variability because it makes things interesting, but sometimes variability can cause problems. Consider the diameters of the cylinders in the engine of your car, which need to be the same. In this case, you want to minimize variability.

Because variability is always with us, we must learn to describe it and work with it rather than ignore it. Where does it come from? How can we adequately describe the important features of variable measurements? These are key questions in statistics and the focus of this activity.

Question

How can we succinctly describe the key features of a set of measurements, such as the shape of the data distribution, the center of the data, the variability of the data, and any "unusual" data points?

Objectives

In this activity, you practice describing sets of measurements in the face of variability, which is present in virtually all measurement problems. If you represent the data graphically, you can see the distribution's symmetry or skewness, its clusters and gaps. If the data are presented numerically, you can compute measures of center and spread. You can find unusual observations—called *outliers*—using appropriate graphs or suitable calculations. Ultimately, you should see that accurate measurements are not easy to make and that any measurement system, no matter how precise, will produce variable measurements at some level. You will also practice using graphical and numerical techniques to describe variation. Following are two additional important lessons.

- Different measurement settings are associated with different sources of variation. Sometimes the variation is "real"—the things you study are intrinsically different. Other times, the measurement itself produces variation. Often, both effects are present at once.
- Sometimes, if you can identify a major source of variation, you can improve the measurement system or at least *standardize* it so that you can compare measurements fairly.

Prerequisites

You must be familiar with plotting data (including box plots with outliers) and calculating numerical summaries of data (mean, median, interquartile range, and standard deviation) before attempting this activity.

Activity

1. Getting ready

   Look out the window. Describe three instances of variability in what you see. Does the variability strike you as good or not so good? What seems to be the cause (or causes) of the variability?

2. Collecting data

   Collect data for one or more of the following as directed by the instructor.

   a. Count the dollar value of the change you currently have with you.

   b. Using the measuring equipment supplied by the instructor, measure the diameter of a tennis ball, to the nearest millimeter.

   c. Using the tape measure provided, measure the circumference of your head, to the nearest millimeter.

   d. Using only a ruler, measure the thickness of a single page of your statistics book, to the nearest thousandth of a millimeter.

   Give the measurement(s) you obtained to the instructor; he or she will provide you with the set of measurements for the class.

3. Analyzing the data

   For each of the settings chosen from the list in step 2, follow these instructions:

   a. Construct a plot of these measurements that shows the shape of the distribution. Describe the shape of the distribution.

   b. Construct a box plot for the set of measurements. Describe any interesting features of this box plot. Are there any outliers shown by the interquartile range rule used in box plots?

   c. Find the mean of the measurements and compare it with the median found for the box plot. Explain why these differ (if, indeed, they do).

   d. Find the standard deviation of the measurements. Identify any observations that are more than 2 standard deviations away from the mean. Are these "unusual" observations the same ones identified by the box plot? Does the standard deviation appear to be a reasonable measure of variability for these data?

   e. Discuss the sources of variation for these measurements.

Wrap-Up

Sometimes a measurement device or system causes variation by producing different results for the same unit. This is true, for example, when using a balance to measure the weight of an object. Two people using the same balance can arrive at slightly different weights for the same object. In fact, one person may come up with differing weights for the same object measured repeatedly over time.

But sometimes variation in measurements occurs because the objects being measured are actually different, as would be true if you counted the number of chairs per classroom in your school. Classrooms have differing numbers of chairs, but, presumably, they can be accurately counted within each room.

Both types of variation affect most measurements, but often one of them predominates. It is important to recognize these differing sources of variation if you want to deal appropriately with variation—or improve the measuring system.

1. Referring back to the measurement problems in this activity, identify the primary source of variation for each problem as resulting from variation among the elements or from variation within the measurement system. If both are major contributors, state this as well.
2. For each of the measurement problems presented in this activity, discuss ways of reducing the variation if you were to repeat these activities.
3. Redo one of the measurement activities, making use of the variation-reduction suggestions from step 2. Did your suggestions work?

Extensions

Sources of variation increase as the system producing the measurements becomes more complicated.

1. How many drops of water will fit on a penny?
   a. Estimate how many drops will fit on a penny. Give your estimate to the instructor.
   b. Using the penny, eyedropper, and cup of water provided, place as many drops of water as possible on the penny. Remember to count while carefully placing the drops. (Teamwork may help here.)
   c. Give your experimental result to the instructor, who will provide you with the data from the class.
   d. Use graphical and numerical methods to analyze these data.
   e. Write a paragraph summarizing the distribution and any other interesting features of these data and commenting on the sources of variation in this experiment.
   f. How would you refine the instructions for this problem to reduce variability?

Assessment Questions

1. Measure the area of your desktop with a meter stick to the closest square centimeter. Do this five separate times.
   a. Comment on the variability among your five measurements.
   b. Comment on the sources of variability for the process of measuring the area of a desktop.
   c. How would you combine the five trials into a single measure of area to report to the rest of the class?

2. To determine how much sleep students get on a typical night, an instructor asked the class to report how many hours they slept last night. The data are shown in the following figure.

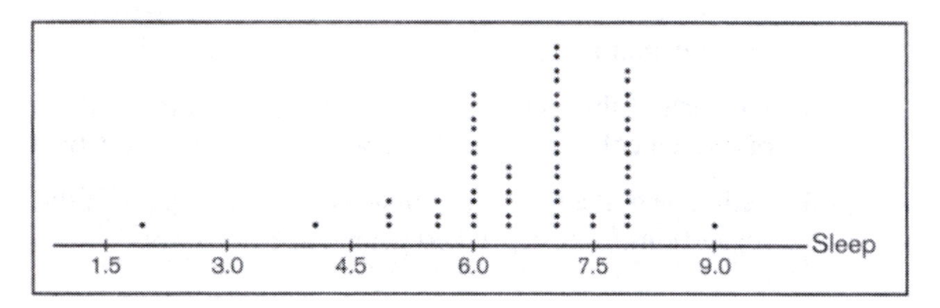

Figure 1

   a. Describe two different sources of variation in these data.
   b. How would you suggest the measurements be made if the goal were to find out how much sleep students get on a typical night?

3. Consider each of the following variables. Do you think it would be better for variability to be high or low? Explain your decision.
   a. Age of trees in a national forest
   b. Diameter of new tires coming off one production line
   c. Scores on an aptitude test given to a large number of job applicants
   d. Daily rainfall

Measurement Bias

Experiencing measurement bias.

Suppose you want to find out how heavy your backpack is. You could measure it by setting it on a bathroom scale, but some bathroom scales are notoriously bad—each individual scale tends to read too high or too low. They have *measurement bias.* You could, however, weigh yourself with the backpack and without it and then subtract to get the pack's weight. In that case—if the scale were really off by a fixed amount—each measurement would have been biased but the final result would not be.

Question

What is measurement bias, and how can we detect it? What useful information can we get from biased measurements?

Objectives

In this activity, you will see that measurement systems are often subject to measurement bias. Your awareness of potential bias in data that you analyze should be heightened as you think about possible sources of bias. Bias is a property of the measurement system; you cannot reduce it simply by taking more measurements. The only way to measure the bias is to compare the measurements with an independent source of "truth" outside the measurement system you are using.

However, *comparisons* between two sets of measurements may be little affected if the bias is the same in each.

Prerequisites

You should have some knowledge of basic data displays, such as dot plots and stem plots, and basic numerical summaries of center and spread, such as the median and interquartile range. You should also have experience describing distributions of data.

Activity

1. Collecting the data

   Look carefully at the string (marked A) the instructor is holding out straight. Without using any measuring instruments (except your eyes), estimate the length of the string to the nearest whole inch.

   Length of string A = ______

   The instructor will collect the estimated string lengths and give you the data for your class. You will use the data in the next part of the activity.

2. Describing the data graphically

   a. Make at least two different plots of the data on string length.

   b. Describe the plots of the data in terms of symmetry or skewness of the distribution; clusters and gaps that might be present; and outliers that might be present, including a possible reason for the outliers.

3. Describing the data numerically

   a. Compute the following numerical summaries of the data: mean, median, standard deviation, and interquartile range.

   b. Which of these measures seems to provide the best description of center? Why?

   c. Which of these measures seems to provide the best description of variability? Why?

4. Collecting and summarizing another set of data

   a. The instructor is now holding another string, string B. As before, estimate the length of this string to the nearest whole inch.

   Length of string B = ______

   The instructor will collect the estimated string lengths and give you the data for your class.

   b. Describe the data for string B using the graphical and numerical techniques you found most useful in the analysis of data from string A.

5. Making comparisons

   From your analysis of the two sets of data, decide which is the longer string. How much longer do you estimate it is?

6. Determining the bias

   a. The instructor will provide you with the correct lengths for each string. Plot the correct values on the plots of the data made previously. What do you see?

   b. It is likely that the true value is not at the center of the data display. This discrepancy between the center of the measurements and the true value is called *bias.* Bias is a property of the measurement *system,* not of an individual person making an estimate. Does the "system" of estimating string lengths appear to be biased? What factors might be causing the bias?

   c. What was the effect of the bias on your answer to step 5? That is, is your estimate of which string is longer and by how much inaccurate because of the bias? Why is this the case?

Wrap-Up

1. Select another activity in which bias may affect the measurements. Collect data within the class or outside the class, as directed by the instructor.

   Possible choices for measurements that are likely to be biased include the following:

   - Estimated height of a doorknob as viewed from across the room
   - Estimated length of the hall outside the classroom
   - Estimated distance to some object across a parking lot or lawn
   - Perceived minute ("Tell me when 1 minute passes")
   - Estimated circumference of your head

   For the set of data you collect:

   a. Analyze it by constructing appropriate plots and numerical summaries
   b. Describe the key features of the distribution in words
   c. Describe what factors might contribute to any bias
   d. Plan and carry out a method for assessing the amount of bias

2. Write a brief summary of what you learned in this activity about measurement bias.

Extensions

1. Find a printed article using data that are subject to a bias that could have a dramatic effect on the conclusions reached in the article. Summarize the conclusions in the article that are based on data, discuss possible biasing factors in the way the data were collected or analyzed, and explain how the bias in the data might affect the conclusion. (Be sure to separate bias in the data from biased reporting of the conclusions for other reasons.)

2. Testing laboratories, such as those that test blood samples or drinking water, are concerned about keeping their measuring equipment accurate. Contact a testing laboratory and ask them about their procedures for reducing bias in the measurements they produce. Give a brief report to the class.

Assessment Questions

1. Working in small groups, measure the area of a tabletop or a wall of the classroom. Of the many ways to do the task, your group should agree on two methods and make a series of measurements of the area in question using each method. There will be variability and bias in the results for each data set. Analyze your results. Your analysis should include a description of the data sets and a statement about variability and bias in each. Which measurement method would you use if you were to do the project again? Are the potential biases large enough to have a serious effect on the result, say, if you were to buy paint to paint the wall or make a glass top for the table?

2. Find an article in the media to critique for possible bias in the measurements used. The critique should include a discussion of possible factors causing the bias and the possible effect of the bias on the conclusions of the article. (Such articles are attached for your use.)

Lead Tests Deemed Erroneous

The results add weight to fears that faulty testing has cost millions of dollars to clean up housing projects.

BY DONYA CURRIE
Sun Staff Writer

Lead-based paint is less of a threat in Gainesville public housing projects than earlier tests reported.

The results received Tuesday from an independent laboratory add weight to officials' fears that faulty testing has cost millions of dollars to clean up housing projects where lead was not a health threat.

Karen Godley, director of maintenance for the Gainesville Housing Authority, said testing done this year by two different companies showed erroneously high levels of lead in several apartment complexes across the city. The problem—a hand-held testing device now under investigation by the federal government for misleading officials in two ways: some tests have shown high levels of lead where the substance posed no health threat; other tests may have failed to detect illegal levels that endangered residents.

Congress banned lead-based paint in 1978 because, if ingested, toxic levels can cause brain damage and learning disabilities in children. Lead testing has become a multimillion dollar industry nationwide, with most money going to companies that garner cleanup contracts of up to $2 million for even a mid-sized housing project.

Gainesville officials contracted with the Miami-based Accutest in 1993 to conduct random tests of city housing projects. When a local company followed up by testing 12 different apartments, almost every result was different. A third test, done this month using paint chip samples instead of a less-expensive, hand-held device, showed lower levels of lead than first detected.

"Before we spent millions or hundreds of thousands of dollars abating lead, we wanted to make sure the results were accurate," said Godley, who expects to replace 100 apartment doors containing high levels of lead.

Officials with the Federal Department of Housing and Urban Development are in the process of rewriting rules for lead testing at public housing projects. In the meantime, Florida housing officials have called for a statewide investigation of test results that may either be putting residents at risk or funneling public money to unnecessary cleanups.

Jim Walker, special assistant in HUD's Jacksonville office, expects Florida officials to take action soon by advising the state's 81 public housing authorities on how to ensure accurate testing. Stories published in The Orlando Sentinel in early April showed Accutest had problems at several housing authorities, reporting high levels of lead where the substance was not a threat and possibly giving cleanup contracts to a related company.

Accutest owner David Mingus denies the charges and came to Gainesville two weeks ago to help with the paint-chip testing. He was not available for comment Tuesday but told The Sun in an earlier interview that he stands by the accuracy of his company's tests.

Godley said she hopes the new HUD guidelines will require only paint-chip testing instead of tests done using a hand-held X-ray device. Housing authorities will pay more for lab analysis of paint chips scraped from apartment walls and doors, but the test results will be accurate, she said.

"I think most of what we're proving right here is that the machine is not reliable," Godley said. "I think the company (Accutest) followed the guidelines. The problem is with the machine."

(continued)

HUD officials are considering changing requirements for lead testers. Florida is one of many states that does not require certification for testers. To enter the business, a person need only purchase a $10,000-$20,000 testing machine and complete a week-long course from the manufacturer.
Source: Gainesville Sun, April 27, 1994.

Instrument Flaw Explains Mystery in Color Vision

NEW YORK (AP)—For 45 years, scientists thought that people's color vision varied according to the seasons. And it left them baffled.

Now science has come up with an answer to one of nature's enduring mysteries. The instrument used for the test in 1948 was flawed.

A study in today's issue of the journal *Nature* found that the instrument's readings are affected by room temperature. So if a testing room got warmer and cooler throughout the year, the device would seem to indicate that people's color vision varied with the seasons, researchers said.
Source: The News-Times (Danbury, CT), June 10, 1993.

3. Optical illusions are related to bias. There are many demonstrations of optical illusions that help prove this point, and some of them are well suited to studies of bias in a measurement process or device. Here is an activity on optical illusion that you might want to try. Collect data from the device explained here and then write a report on the results.

   One simple device that works well to show bias in our visual judgments is illustrated on the following pages. The goal is to make the line with arrowtails at the ends equal in length to the line with arrowheads at the ends. The head line is of fixed length, drawn on the cardboard in advance. The line with arrowtails slides out from behind the front cardboard so that the person making the "measurement" is free to stop it at any point. The length of the fixed, arrowhead line is the "truth." It is the job of the participant to make the length of the line with arrowtails as close to truth as possible.

The size of the device can change depending on how it is to be used. For small groups, an 8-inch square background is large enough, but class demonstration may require a larger one. Rules for sliding out the back arrow can be formulated by the class. Following are two that work well.

a. Slide the arrow out until you think the lengths match and then stop, with no "backsliding" allowed.

b. Slide the arrow back and forth at will until you think the lengths match.

You may work in groups to collect data, but each student should repeat the procedure three or four times so that you can assess individual biases and variability. Plot the measurements of the length of the "arrow-tail" line on a dot plot for each student; then combine the data to make a plot for each group. Discuss bias in the measurement process.

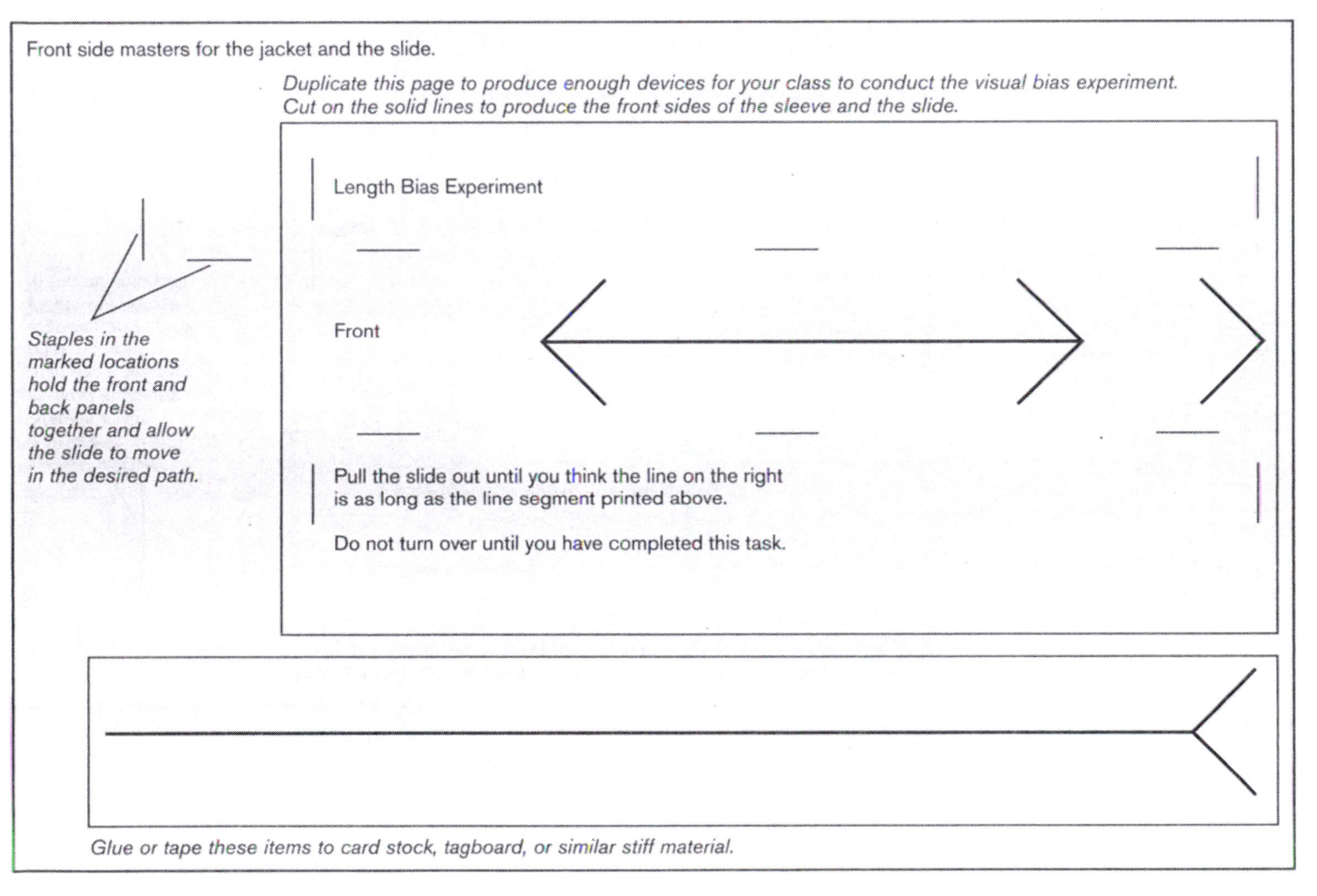

Figure 1: Length bias experiment

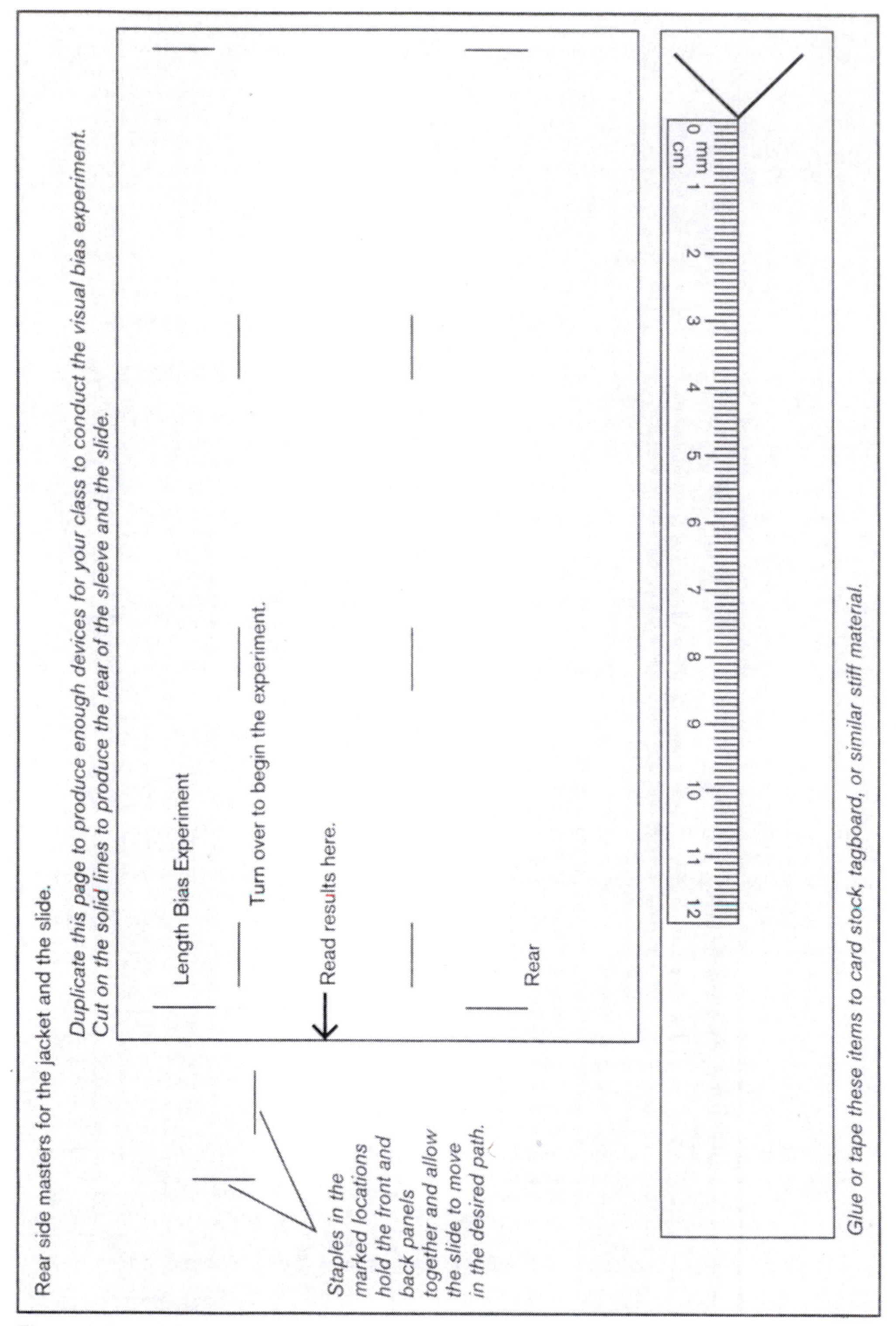

Figure 2: Length bias experiment

Let Us Count

Variation due to the process of measurement.

History repeats itself, but never exactly. An athlete will not run a mile in the identical time twice; you may not get the same number of French fries in two different orders; two scoops of ice cream will not be exactly the same. This "variability" is present in all repeated actions. *Process variability* refers to the way outcomes vary as a result of the process of measurement—a major concern in industry. The main objective in process control is to figure out what causes the variation—and to control it.

Questions

Which process of measurement is better? How do differences in measurement processes affect the variability in the outcomes?

Objectives

This activity illustrates the presence of variability in a process and analyzes this variability. You will learn how to compare the results of two measurement processes, as well as the patterns and sources of variation in each process.

Prerequisites

You should know how to construct and interpret stem and leaf diagrams, and you should be familiar with means and standard deviations as measures of center and spread.

Activity

Suppose you manage an ice-cream parlor. Employees use different techniques to scoop out half-cup servings. You notice that these half-cup servings vary in size. As a manager, you need to find a way of scooping—a measurement technique—that varies as little as possible so that you can plan and run your business better. Unfortunately, rather than using ice cream in this classroom experiment, you'll be using pasta shells instead.

1. Collecting the data

   The class will be divided into an even number of teams consisting of at least two students each. The teams will then be separated into two groups.

   Within the first group, each team pours shells into a half-cup measure and counts the number of shells.

   a. One person pours shells into the half-cup measure until he or she thinks it is full.

   b. A second person then counts the number of shells and records it, without informing the first person of this number.

   c. Each team repeats the experiment five times and computes the mean ($\bar{x}$), the standard deviation (s), and the variance (s^2) for its measurements.

   In the second group, students use a larger transparent plastic measure to measure a half cup of shells.

   a. One person in the team pours shells into the large measuring vessel until he or she thinks it has reached the half-cup mark.

   b. A second person then counts the number of shells and records it, without informing the first person of this number.

   c. Again, each team repeats the experiment five times and computes the mean ($\bar{x}$), the standard deviation (s), and the variance (s^2) for its measurements.

2. Making plots and boundaries

   a. Construct stem and leaf diagrams of the mean counts of the teams for each group and combine these two plots in a back-to-back stem and leaf diagram.

   b. Construct stem and leaf diagrams for the standard deviations of the teams for each group and combine these in a back-to-back stem and leaf diagram.

   c. Compute the two boundaries for these means, $\bar{\bar{x}} - (2s/\sqrt{n})$, and $\bar{\bar{x}} + (2s/\sqrt{n})$ for each group, where $n = 5$, the number of measurements from each team, $\bar{\bar{x}} =$ the overall mean for the group, and $s^2 =$ the average of the variances for that group. We also call this quantity s^2 the *mean squared error.*

   d. How many of the sample means for each group fall outside these boundaries?

3. Analyzing the results
   a. What do the stem and leaf diagrams of the sample means and standard deviations indicate about the variation within each group and between the groups? Do both processes vary by about the same amount? Are there differences in the patterns of variation?
   b. Identify the different sources of variation. Which would you consider common cause variation and which special cause variation?

> Note: We often divide variability in processes into two types:
>
> - *Common cause* variation is part of the system or process and affects everyone in the system.
> - *Special cause* variation either is not part of the system or process all the time or does not affect everyone in the system.

   c. Compare the overall means and the boundaries for the two groups. Which of these two methods would you recommend and why?

Wrap-Up

This type of analysis is often used in industry to understand the processes used to manufacture products or provide various services. This activity looked at a simple process, investigated the variability, and checked for unusual observations. There is a theoretical basis for using this approach. Answering the following questions should help you understand this basis.

1. Why did we use the boundaries of $\bar{\bar{x}} - (2s/\sqrt{n})$ and $\bar{\bar{x}} + (2s/\sqrt{n})$ to check for unusual team means? Should we expect many team means to fall outside these limits? What is the approximate probability of a sample mean falling outside the boundaries that you have used?

2. From knowledge of the behavior of sample means, can you explain why the distance between the boundaries will get smaller for larger sample sizes? What does that imply about the variability of the sample means?

Extensions

Instead of counting the number of shells, would weighing the shells result in a process with less variation? Depending on the scales and weights available to you, take repeated samples of the shells of a specified weight, count the number of shells, and construct a stem and leaf diagram and a control chart. Note that here you have only one measurement process. Assume that you are the quality manager. Write a report to the production manager that outlines the results of your experiment and the extension activity and that recommends a measurement process.

Assessment Questions

1. In attempting to control a process, why is it important to systematically check both the center and the variability of measurements of the process?

2. In this activity, you averaged five measurements and checked the averages against the boundaries. What is the advantage of working with the small samples of 5 as opposed to simply checking individual measurements against a boundary? Which method would tend to show "out of control" cases more often?

3. If the boundaries given in this activity are followed, a costly process might be shut down if a sample mean lies beyond $2s/\sqrt{n}$ (where s is defined in "Activity" step 2c) from the overall mean. Some quality control plans replace the $2s/\sqrt{n}$ with $3s/\sqrt{n}$. Discuss the pros and cons of using $3s/\sqrt{n}$ boundaries instead of $2s/\sqrt{n}$.

Matching Descriptions to Scatter Plots

Making the correspondence between scatter plots and statistics (regression line and r).

Table 1 and the scatter plot in Figure 1 show marriage and divorce rates (per 1000 people per year) for 14 countries. The correlation coefficient for these data is $r = 0.597$, and the equation of the least-squares regression line is $y = -0.7 + 0.5x$.

Marriage Rate per 1000 People per Year	Divorce Rate per 1000 People per Year	Marriage Rate per 1000 People per Year	Divorce Rate per 1000 People per Year
5.6	2.0	6.1	1.9
6.0	3.0	4.9	2.2
5.1	2.9	6.8	1.3
5.0	1.9	5.2	2.2
6.7	2.0	6.8	2.0
6.3	2.4	6.1	2.9
5.4	0.4	9.7	4.8

Table 1: Marriage and divorce rates

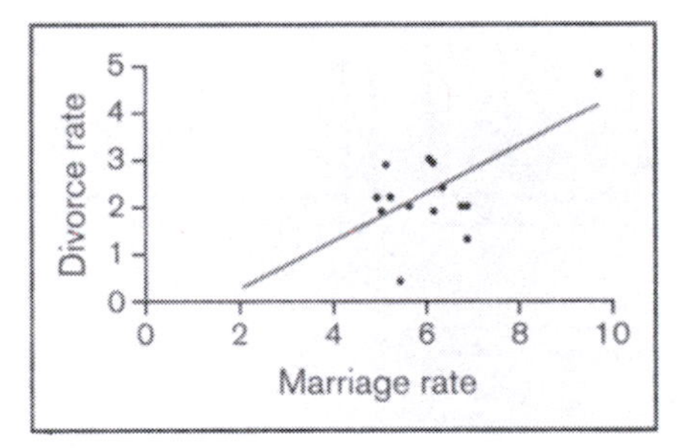

Figure 1: Scatter plot of marriage and divorce rates

Question

On the basis of the regression line, predict the divorce rate for a country with a marriage rate of 8 per 1000. How much conviction do you have in this prediction?

Objectives

In this activity, you will learn how one point can influence the correlation coefficient and regression line.

Prerequisites

You can do this activity as soon as you have learned about regression and that the correlation coefficient measures spread about the least-squares regression line.

Activity

1. Match each of the five scatter plots to the description of its regression line and correlation coefficient. The scales on the axes of the scatter plots are the same.
   a. $r = 0.83, y = -2.1 + 1.4x$
   b. $r = -0.31, y = 7.8 - 0.5x$
   c. $r = 0.96, y = -2.1 + 1.4x$
   d. $r = -0.83, y = 11.8 - 1.4x$
   e. $r = 0.41, y = -1.4 + 1.4x$

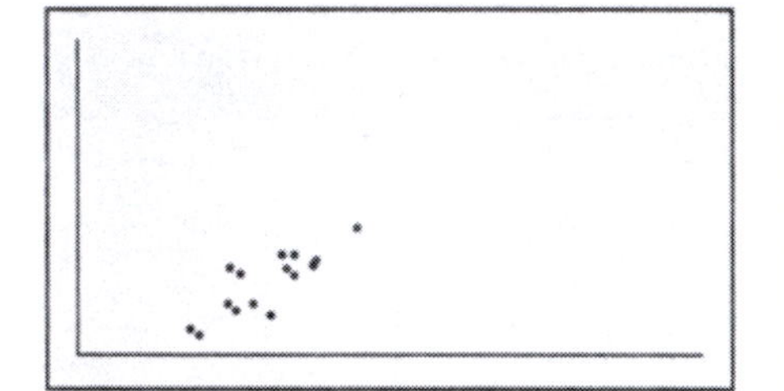

Scatter plot 1

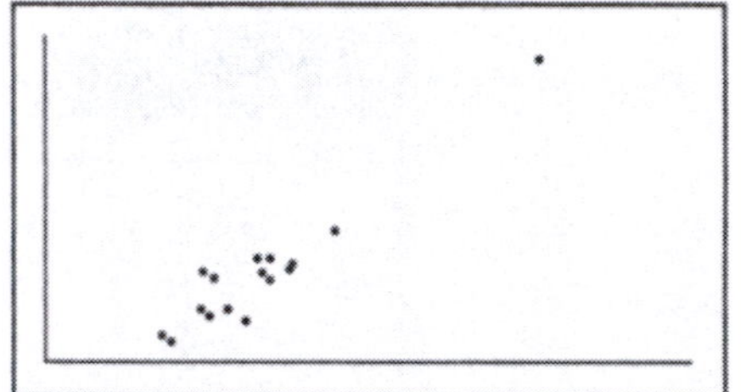

Scatter plot 2

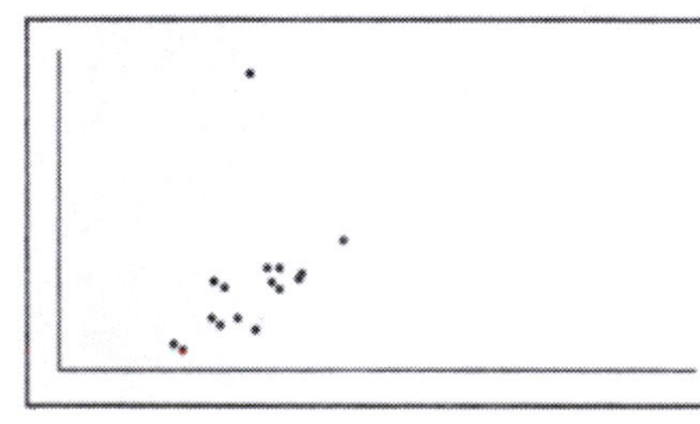

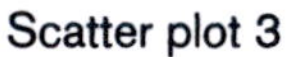

Scatter plot 3

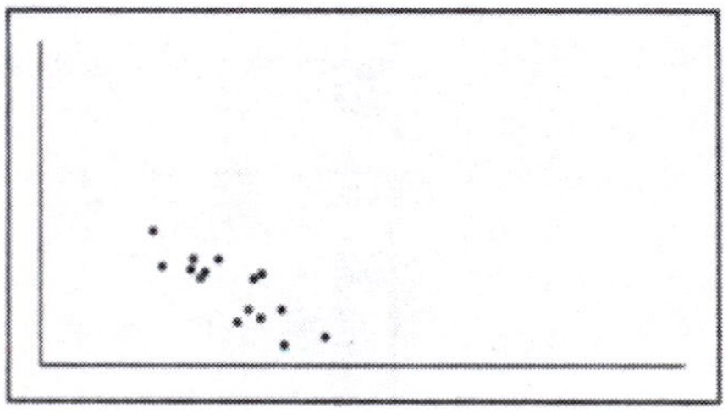

Scatter plot 4

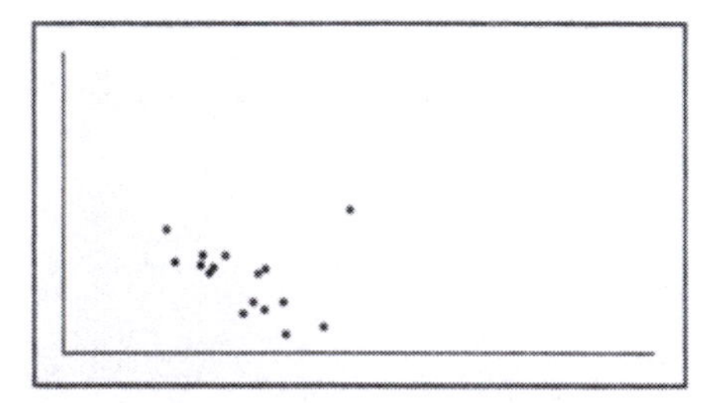

Scatter plot 5

2. For the nine points on the following scatter plot, $r \approx 0.71$, $r^2 = 0.5$, and the equation of the least-squares regression line is $y = 4.00 + 1.00x$.

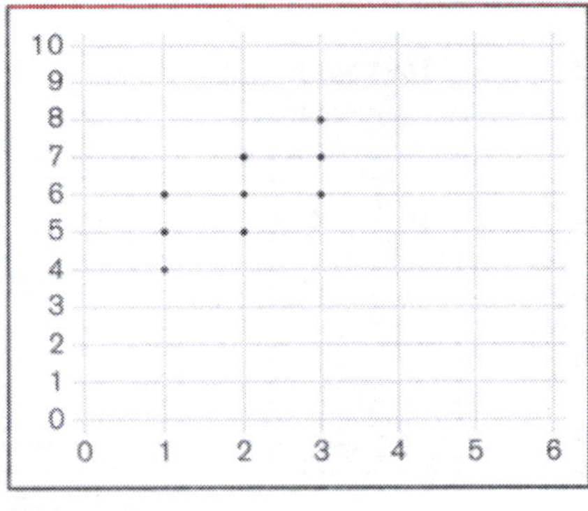

Figure 2

A tenth point is added to the original nine. Match each of the following points with the correlation coefficient that would result if that point were added. Do not calculate the new correlation coefficient but rather reason out which r must go with each point. Correlation coefficients are as follows: −0.84, −0.70, 0.02, 0.22, 0.71, 0.73, 0.96.

a. (3, 7)
b. (2, 6)
c. (10, 0)
d. (10, 6)
e. (10, 14)
f. (100, 0)
g. (100, 6)

Wrap-Up

1. Look again at the scatter plot in Figure 1. The country at point (9.7, 4.8) is the United States. The other 13 are European countries. Compute the correlation coefficient and the regression equation when the United States is removed from the data. What is your conclusion?

2. Where would you place a new point on a scatter plot to make the correlation coefficient as close to 1 as possible? As close to 0 as possible?

Extension

Analyze the four data sets (from Anscombe, 1973) in the following table. What do they have in common? Why are they of interest? What do they illustrate?

Data Set 1		Data Set 2		Data Set 3		Data Set 4	
x	y	x	y	x	y	x	y
10	8.04	10	9.14	10	7.46	8	6.58
8	6.95	8	8.14	8	6.77	8	5.76
13	7.58	13	8.74	13	12.74	8	7.71
9	8.81	9	8.77	9	7.11	8	8.84
11	8.33	11	9.26	11	7.81	8	8.47
14	9.96	14	8.10	14	8.84	8	7.04
6	7.24	6	6.13	6	6.08	8	5.25
4	4.26	4	3.10	4	5.39	19	12.50
12	10.84	12	9.13	12	8.15	8	5.56
7	4.82	7	7.26	7	6.42	8	7.91
5	5.68	5	4.74	5	5.73	8	6.89

Table 2

Assessment Questions

Draw a scatter plot using a set of points that have negative correlation when taken together but positive correlation when one point is removed. Circle the influential point.

The Regression Effect

Find out about the regression effect.

A mathematics supervisor in a large U.S. city received a grant to improve mathematics education. She tested all students and placed those with the lowest achievement scores in a special program. After a year, she retested all of the students and was gratified to see that the students in the special program improved in comparison with the rest of the students.

A few years ago, a school in New Jersey tested all of its fourth graders to select students for a program for the gifted. Two years later the students were retested, and the school was shocked to find that the scores of the gifted group had dropped in comparison with the rest of the students.

Question

Can we conclude that the remedial program worked and that the program for the gifted was detrimental?

Objectives

After completing this activity, you will understand the regression effect (sometimes called *regression toward the mean*) and will know how to recognize where it might occur.

Prerequisites

You should be able to read and interpret scatter plots. It is best if you have already studied correlation and regression.

Activity

1. Take the following test, individually.

> **MONEY TEST**
>
> This is a closed-wallet, multiple-choice test. If you are not sure of an answer, make your best estimate. Don't leave any questions blank.
>
> **I.** On the back of a nickel is
> **a.** Monticello
> **b.** The Jefferson Memorial
>
> **II.** On the back of a $2 bill is
> **a.** Signers of the Declaration
> **b.** Independence Hall
>
> **III.** On the front of a $500 bill is
> **a.** Madison
> **b.** McKinley

Your instructor will provide the answers.

2. After you have been told to do so, take the make-up test.

> **MONEY MAKE-UP TEST**
>
> This is a closed-wallet, multiple-choice test. If you are not sure of an answer, make your best estimate. Don't leave any questions blank.
>
> **I.** On the front of a $20 bill is
> **a.** Jefferson
> **b.** Jackson
>
> **II.** On a dollar bill, Washington is looking to his
> **a.** Left
> **b.** Right
>
> **III.** On the front of a $1000 bill is
> **a.** Cleveland
> **b.** Wilson

Your instructor will supply the answers.

3. What was the average score on the first test for the "remedial" students in your class (those students who missed two or three questions)? What was the average score on the second test for the "remedial" students?
4. What average score did the "star" students (those who got all three questions right on the money test) get on the make-up test?
5. Explain why the scores of the "remedial" students tended to go up and the scores of the "star" students tended to go down. This phenomenon is called the *regression effect,* or *regression toward the mean.*
6. The following scatter plot shows the heights of 1078 fathers and their grown sons in about 1900 (see Freedman et al., 1991). The average height of all the sons is 69 inches, and the average height of all the fathers is 68 inches. The shape of this scatter plot is typical of two positively correlated variables. Taller fathers tend to have taller sons, and shorter fathers tend to have shorter sons; however, we cannot predict with certainty a son's height based on his father's height. Even if we knew the mother's height, we still could not predict the son's height exactly. Some element of chance is involved.

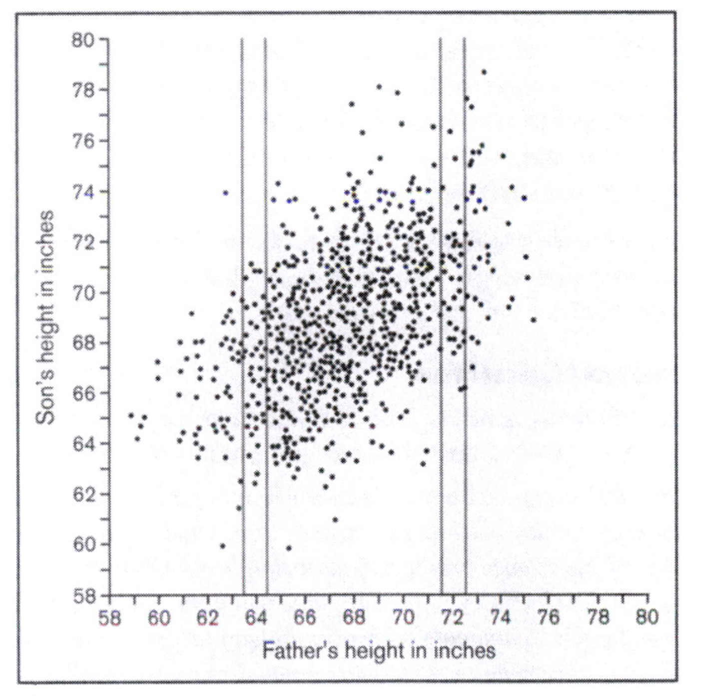

Figure 1

a. Use the vertical strip on the right in Figure 1 to estimate the average height of the sons of 72-inch-tall fathers. Place an "X" at this point on the graph.
b. Use the vertical strip on the left in Figure 1 to estimate the average height of the sons of 64-inch-tall fathers. Place an "X" at this point on the graph.

c. Connect the two X's with a line. This line approximates the regression line.

d. Use the regression line to estimate the average height of the sons of 68-inch-tall fathers.

e. Explain how the father and son data exhibit the regression effect.

Wrap-Up

1. How does the regression effect apply to the two situations at the beginning of this lesson—the one with the mathematics supervisor with a special program for low achievers and the one about a gifted program in a New Jersey school?

2. Explain why "regression toward the mean" is a logical name.

Extensions

1. Your instructor will supply the scores from the first and second tests in a college statistics class. Make a scatter plot of these scores. Does this scatter plot exhibit perfect correlation? Identify which points were for the students who did well on the first test and which were for the students who did poorly. What happened to these students on the second test compared with the other students?

2. Suppose we hold an interspecies Olympics 100-meter dash. The contestants are a turtle, a snail, an armadillo, a human toddler, an adult human sprinter, a horse, and a cheetah. Each participant gets to run the race twice. Comparing the first and second run, will the times show the regression effect? Why or why not? What makes this different from (or the same as) the test-retest situations we have been studying?

3. Find some data to illustrate that stocks on the New York Stock Exchange exhibit the regression effect. Does this mean that a person should avoid buying the stocks that did well the previous year?

Assessment Questions

1. The following quote is from Francis Galton's *Memories of My Life* (London: Methuen, 1908). Galton first recognized the regression effect.

   The following question had been much in my mind. How is it possible for a population to remain alike in its features, as a whole, during many successive generations, if the average produce of each couple resemble their parents? Their children are not alike but vary: therefore some would be taller, some shorter than their average height; so among the issue of a gigantic couple there would be usually some children more gigantic still. Conversely as the very small couples. But from what I could thus far find, parents had issue less exceptional than themselves. I was very desirous of ascertaining the facts of the case.

Translate Galton's statement into more modern English. What assumption is Galton making that isn't true? Write an explanation to Galton.

2. From almanacs for 2 consecutive years, gather individual batting averages for major league baseball (MLB) players. What does the regression effect mean in this context? Do the batting records exhibit the regression effect? How can you tell without analyzing data for all of the players?

3. Psychologists once recommended that instructors in a flight school praise any exceptionally good execution of a flight maneuver by a trainee. After trying this approach for some time, the instructors reported that positive reinforcement did not work. The trainees who had been praised for a good maneuver usually performed less well on the next try. Can you suggest a probable cause for this result? (See Kahneman, Slovic, and Tversky, 1982, pp. 66–68.)

Leonardo's Model Bodies

Looking at correlation between the sizes of different body parts.

Leonardo da Vinci (1452–1519), the great painter and scientist of the late 15th and early 16th centuries, studied the human body and provided detailed accounts of the relationships among various body parts. Leonardo's work on how the size of one part is related to that of another was for the purpose of instructing painters on how to paint the human body, but some of these relationships are of practical interest as well. (Shirtmakers might be interested in the relationship between arm length and circumference of the neck, for example.) In this activity we will investigate the accuracy of some of Leonardo's pronouncements by modeling relationships between the sizes of various body parts.

Question

How can we use a mathematical equation to represent the relationship between two variables? And how can we use data to get that equation right?

Objective

In this activity, you will examine the relationship between two variables by fitting a least-squares regression model to paired data and interpreting the slope of the resulting straight line.

Prerequisites

You should be able to make and understand a scatter plot, and given the equation for a line, you should be able to draw the line on that scatter plot. In other words, you must know about slope and intercept.

Activity

1. Read the article on Leonardo's view of the human body (pages 50–52).
2. Of the many relationships mentioned in the article, choose at least two to investigate. For example, you might study the relationship between height and arm span and between hand length and arm length. Clearly state the relationships you are investigating.
3. List all of the variables you need to investigate the relationships you chose in the problem statement. Decide how to measure these parts on any one person.
4. Select a group of people (at least 15) on which to make the measurements. Collect the necessary data.
5. Construct scatter plots to compare the variables in all of the relationships you are studying. For the examples cited earlier, one scatter plot would show height versus arm span and another would display hand length versus arm length. Describe any patterns you see in the data.
6. If appropriate, find the equation of a simple linear regression model that relates one of the paired variables to the other. Draw the line on the scatter plot and comment on how well it fits.
7. Interpret the data by looking carefully at the two estimated parameters of each model: the slope and the intercept. How do the slopes relate to Leonardo's conjectures? What should happen to the intercepts in most of these models?

Wrap-Up

1. Once a model has been fit to data from one group of people and analysis has suggested that it fits well, verify that the model works on another small group of people.
2. Plan and carry out another investigation of this type. If the result of step 1 shows some problems with the model, plan a new investigation to see if the problems can be surmounted. If the verification shows the model to be working, specify another relationship that you might investigate.

Extensions

1. Investigate what happens to the models (linear equations) if the roles of x and y are reversed. For example, if arm length (x) is used to predict hand length (y), what will happen to the equation if hand length (now x) is used to predict arm length (now y)? Will a similar result hold when interchanging arm span and height?
2. Investigate the linear models more carefully by plotting the residuals and studying any patterns that might emerge there. Do the residuals suggest that you should investigate a more complex model?
3. Calculate correlation coefficients for each of the scatter plots you made during the main activity. Interpret these measures of the strengths of the linear relationships.

Leonardo's View of the Human Body

The following excerpt is from Martin Kemp's *Leonardo on Painting* (New Haven, CT: Yale University Press, 1989).

On the Measurements of the Human Body

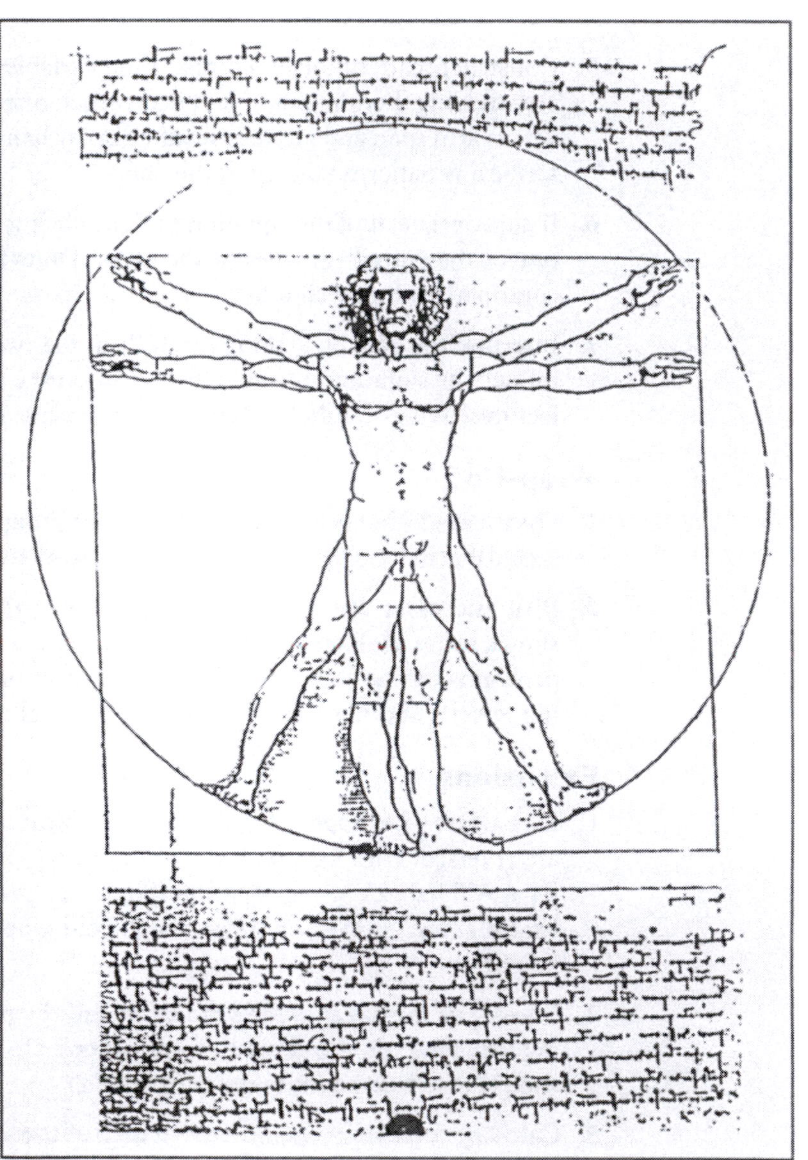

Vitruvius, the architect, has it in his work on architecture that the measurements of man are arranged by nature in the following manner: four fingers make one palm and four palms make one foot; six palms make a cubit; four cubits make a man, and four cubits make one pace; and twenty-four palms make a man; and these measures are those of his buildings.

If you open your legs so that you lower your head by one-fourteenth of your height, and open and raise your arms so that with your longest fingers you touch the level of the top of your head, you should know that the central point between the extremities of the out-stretched limbs will be the navel, and the space which is described by the legs makes an equilateral triangle.

The span to which the man opens his arms is equivalent to his height.

From the start of the hair [i.e., the hairline] to the margin of the bottom of the chin is a tenth of the height of the man; from the bottom of the chin to the top of the head is an eighth of the height of the man; from the top of the breast to the top of the head is a sixth of the man; from the top of the breast to the start of the hair is a seventh part of the whole man; from the nipples to the top of the head is a quarter part of the man; the widest distance across the shoulders contains in itself a quarter part of the man; from the elbow to the tip of the hand will be a fifth part of the man; from this elbow to the edge of the shoulder is an eighth part of this man; the whole hand is a tenth part of the man.

If a man of two *braccia* is small, one of four is too large—the mean being the most praiseworthy. Halfway between two and four comes three. Therefore take a man of three *braccia* and determine his measurements with the rule I give you. And if you should say to me that you might make a mistake and judge someone to be well proportioned who is not, I reply on this point that you must look at many men of three *braccia,* of whom the great majority have limbs in conformity with each other. From one of the most graceful of these take your measurements. The length of the hand is a third of a *braccio* and goes nine times into the man; and correspondingly the face, and from the pit of the throat to the shoulder, and from the shoulder to the nipple, and from one nipple to the other, and from each nipple to the pit of the throat.

- The space between the slit of the mouth and the base of the nose is one-seventh of the face.

- The space from the mouth to below the chin, *cd,* will be a quarter part of the face, and similar to the width of the mouth.

- The space between the chin and below the base of the nose, *ef,* will be a third part of the face, and similar to the nose and the forehead.

- The space between the midpoint of the nose and below the chin, *gh,* will be half the face.

- The space between the upper origin of the nose, where the eyebrows arise, *ik,* to below the chin will be two-thirds of the face.

- The space between the slit of the mouth and above the beginning of the upper part of the chin, that is to say, where the chin ends at its boundary with the lower lip, will be a third part of the space between the parting of the lips and below the chin, and is a twelfth part of the face. From above to below the chin, *mn,* will be a sixth part of the face, and will be a fifty-fourth part of the man.

- From the furthest projection of the chin to the throat, *op,* will be similar to the space between the mouth and below the chin and is a quarter part of the face.

- 

The space from above the throat to its base below, *qr,* will be half the face and the eighteenth part of the man.

- 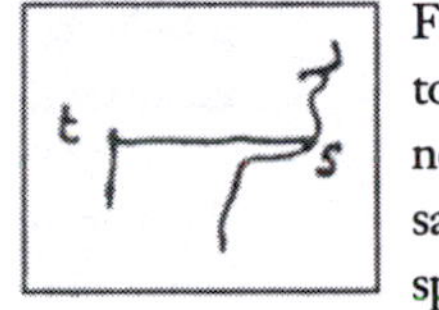

From the chin to behind the neck, *st,* is the same as the space between the mouth and the start of the hair, that is to say three-quarters of the head.

- 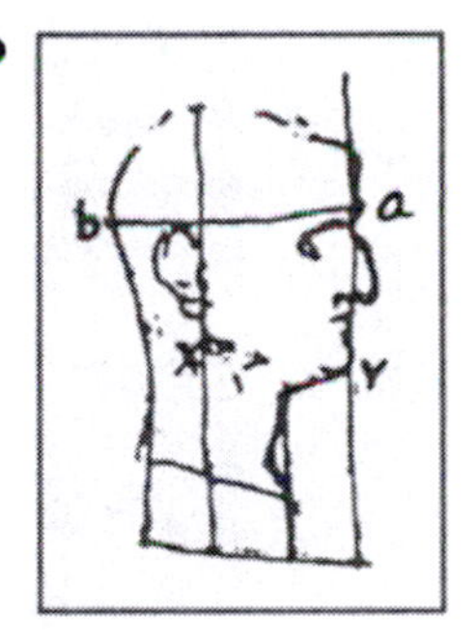

From the chin to the jawbone, *vx,* is half the head, and similar to the width of the neck from the side.

- 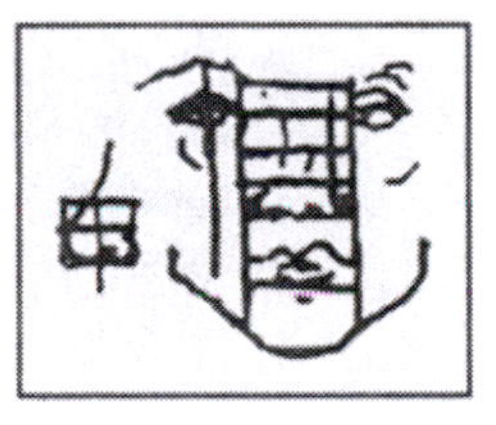
The thickness of the neck goes one and three-quarter times into the space from the eyebrows to the nape of the neck.

- The distance between the centers of the pupils of the eyes is one-third of the face.

- The space between the edges of the eyes towards the ears, that is to say, where the eye ends in the socket which contains it (at its outer corners), will be half the face.
- The greatest width which the face has at the level of the eyes will be equivalent to that between the line of the hair at the front and the slit of the mouth.
- The nose will make two squares, that is to say, the width of the nose at the nostrils goes twice into the length between the tip of the nose and the start of the eyebrows; and, similarly, in profile the distance between the extreme edge of the nostril—where it joins the cheek—and the tip of the nose is the same size as between one nostril and the other seen from the front. If you divide the whole of the length of the nose into four equal parts, that is to say, from the tip to where it joins the eyebrows, you will find that one of these parts fits into the space from above the nostrils to below the tip of the nose, and the upper part fits into the space between the tear duct in the inner corner of the eye and the point where the eyebrows begin; and the two middle parts are of a size equivalent to the eye from the inner to the outer corner.
- The hand to the point at which it joins with the bone of the arm goes four times into the space between the tip of the longest finger and the joint at the shoulder.
- *Ab* goes four times into *ac*, and nine times into *am*. The greatest thickness of the arm between the elbow and the hand goes six times into *am*, and is similar to *rf*. The thickest part of the arm between the shoulder and the elbow goes four times into *cm*, and is similar to *hng*.

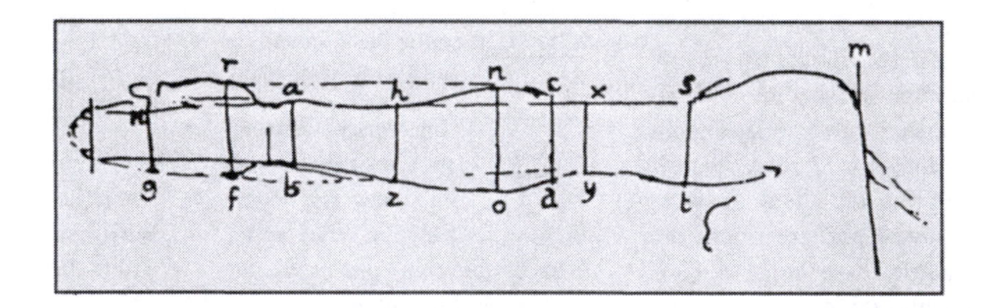

- The least thick part of the arm above the elbow, *xy*, is not the base of a square, but is similar to half of the space *hz*, which is found between the inner joint of the arms and the wrist. The width of the arm at the wrist goes twelve times into the whole of the arm, that is to say, from the tip of the fingers to the shoulder joint, that is to say, three times into the hand and nine times into the arm.
- If a man kneels down he will lose a quarter of his height. When a man kneels down with his hands in front of his breasts, the navel will be at the midpoint of his height and likewise the points of his elbows.

Assessment Questions

1. The plot and statistical summaries for paired data on kneeling height versus height from a sample of 15 students is shown in Figure 1.
   a. Describe the pattern in the scatter plot.
   b. Interpret the slope of the fitted regression line in terms of these measurements.
   c. Would you say that the line fits the data well? Explain.

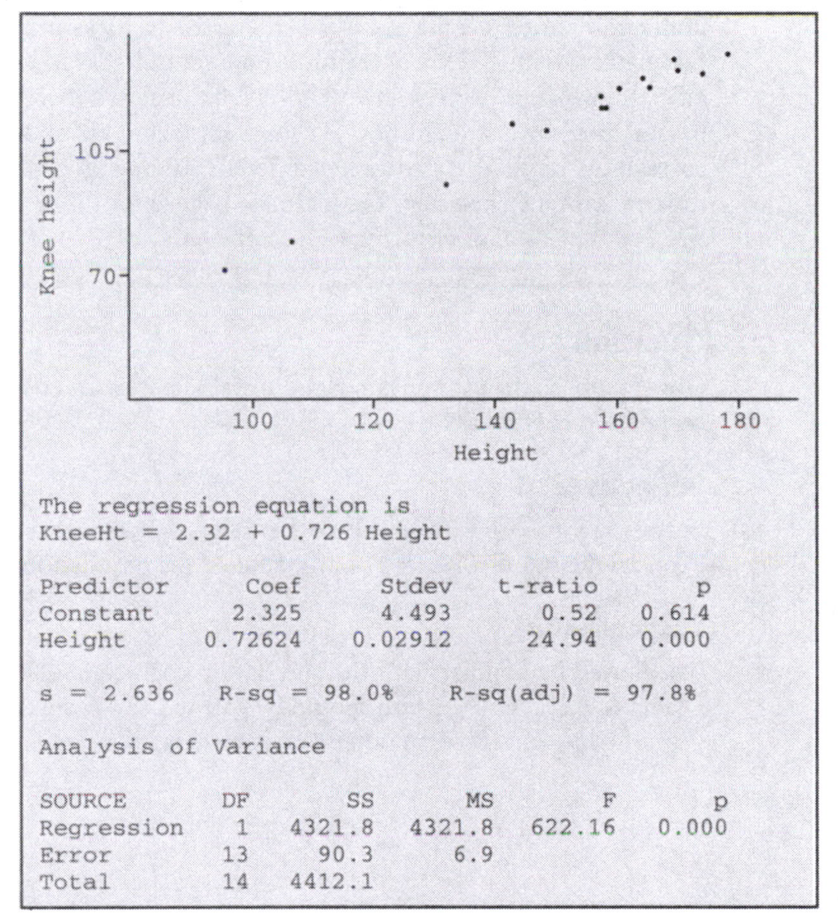

The regression equation is
KneeHt = 2.32 + 0.726 Height

Predictor	Coef	Stdev	t-ratio	p
Constant	2.325	4.493	0.52	0.614
Height	0.72624	0.02912	24.94	0.000

s = 2.636   R-sq = 98.0%   R-sq(adj) = 97.8%

Analysis of Variance

SOURCE	DF	SS	MS	F	p
Regression	1	4321.8	4321.8	622.16	0.000
Error	13	90.3	6.9		
Total	14	4412.1			

Figure 1

2. Some students are writing a report on the accuracy of Leonardo's conjectures. The report stresses where Leonardo seems to be on target and where he seems to be in error. Speculate as to why Leonardo's conjectures might be way off for some of the relationships investigated by these students.

Relating to Correlation

How the correlation coefficient from a sample varies about the true population coefficient.

Is your high-school GPA a good indicator of your college GPA? Is a high cholesterol level related to high fat intake? Are your grades related to the number of hours you study? People are often interested in measuring the relationships between two variables. One of the most popular statistics to use is the correlation coefficient, which measures how well the relationship between the two variables can be described by a straight line. We use a sample to estimate the true (population) correlation coefficient, so we get different data—and a different estimate—each time we sample. How does this estimate behave for different values of the sample size and the population correlation coefficient?

Question

How "good" is the sample correlation coefficient as an estimate of the population correlation coefficient?

Objectives

In this activity, you will learn how the values of sample correlation coefficients depend on the sample size and on the value of the population correlation coefficient.

Prerequisites

You should be familiar with the correlation coefficient. This activity uses Fathom Dynamic Statistics™, so you should be able to use a computer, know how to select from menus, and perform other computer-related tasks.

Activity

This activity is based on a computer simulation that uses Fathom. You will generate samples of two different sizes from populations that have specified correlation coefficients.

Generating the Random Samples

1. Your instructor will provide you with the file **Relating to Correlation.ftm.** Open it. The left-hand part of the screen will look like Figure 1.

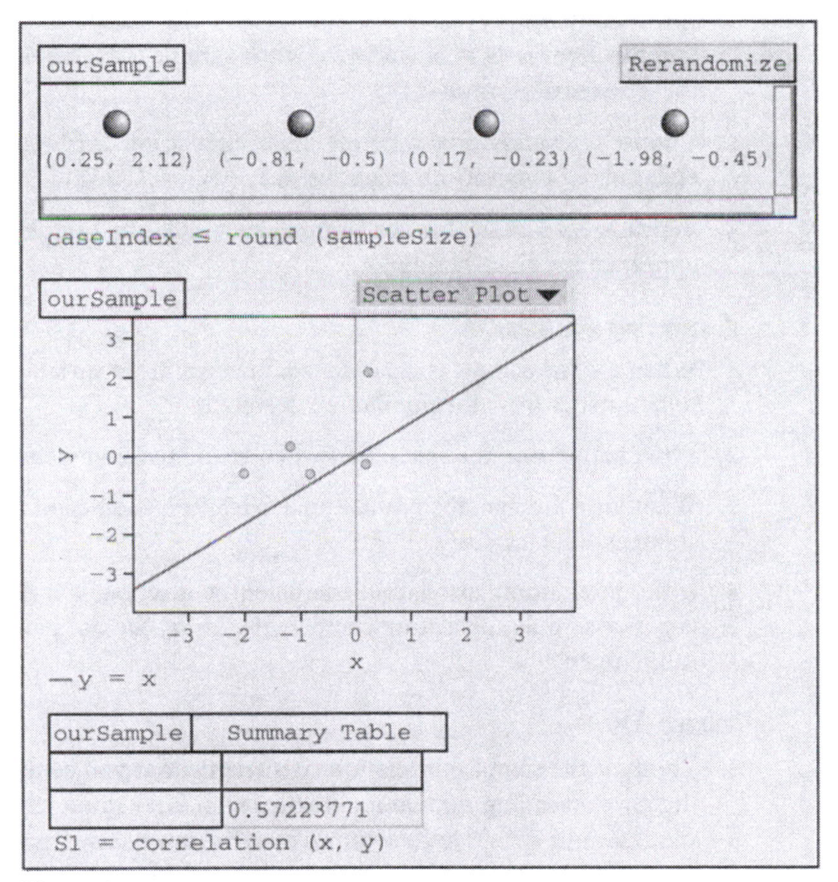

Figure 1

2. Press the **Rerandomize** button in the top box. Note how the five points change in the scatter plot below it. Note also the table at the bottom that shows the correlation coefficient for the sample of five points. This table changes with each rerandomization.

3. To the right of the box with **Rerandomize** are two sliders named **SampleSize** and **TrueCorrelation.** Move the pointers on the sliders to change the values. See how that affects the scatter plot and correlation coefficient.

4. Reset the sliders so that **SampleSize** = **5.0** and **TrueCorrelation** = **0.1**. You can use **Undo** or simply edit the numbers in the slider. Press **Enter** or **Tab** when you're done editing.

Now we want to see how the sample correlation varies among different samples. We'll have Fathom rerandomize automatically, 20 times, and see what the distribution of sample correlations looks like.

5. In the box in the upper right, click the button labeled **Collect More Measures.** Gold balls will fly across the screen, and points will appear in the middle (previously empty) dot plot that shows the sample correlation coefficient, **SampleCorr** for **N** = 5. Notice how the sample correlations are spread around the true correlation of 0.1.
6. Choose four additional different **TrueCorrelations** and repeat step 5. An interesting set of correlations might be {0.1, −0.5, 0.4, 0.75, 0.9}.
7. Repeat steps 5 and 6 but for **SampleSize** = **50**. The sample correlations should appear in the right-hand plot.

Analyzing the Results

1. When are the sample correlation coefficients more variable, that is, more likely to be further from the population correlation?
2. What happens to this variability when we increase the sample size?
3. What does the sign of a population correlation coefficient indicate about how y changes as x changes?
4. If the population correlation coefficient is positive, is it likely that we can get negative sample correlation coefficients? Using the dot plots, discuss when that can happen.

Wrap-Up

1. Think of the sample correlation coefficients that you generated as estimates of the corresponding population coefficients. How good are these estimates for small sample sizes? Do they improve as the sample size increases?
2. Did the answer to question 1 depend on the population correlation coefficient?
3. Write a short summary of the results of your simulation and suggest a "rule of thumb" to get a reliable estimate of r.

Extensions

1. Set **SampleSize** to 50 and look at the scatter plot on the left. Compare the pattern of points in the plots around the $y = x$ line as the population correlation coefficient decreased from 0.9 to 0.1.
2. Repeat step 1 for **SampleSize** = **200**.

3. Explain why you think it's likely to get a sample of 5 observations far from the $y = x$ line when the population correlation coefficient is small (e.g., 0.1).
4. Does the probability of doing so change when the population correlation coefficient increases?

Assessment Questions

1. A newspaper reports that the correlation between two variables is 0.7 and that the result is a highly significant finding. It goes on to say that correlation is based on 10 data points. Do you agree with the newspaper's interpretation? Explain.
2. How would your answer to question 1 change if the sample size were 500 instead of 10?
3. Is it possible to have a negative sample correlation when, in fact, two variables are known to be positively correlated? Explain.

Models, Models, Models . . .

Modeling time series data with lines. Breaking the time series into two pieces.

When we examine the relationship between two quantities, we are often interested in fitting a mathematical function to describe this relationship. Statisticians call these functions *models*. In this activity, you will use data on levels of carbon dioxide (CO_2) in the atmosphere to illustrate fitting a model to data.

The atmospheric levels of CO_2 have been increasing. Many scientists believe that increasing levels of gases such as CO_2 could contribute to global warming, which could lead to disastrous coastal flooding, cause severe droughts, and generally alter the climatic systems under which most of the life on Earth developed.

You can read additional background about the problem and these data on page 62 (Mauna Loa article and graphic).

Question

How has the level of CO_2 changed over time? What mathematical function will best "model" the relationship between the two variables?

Objectives

In this activity, you will learn how to detect when a variable is changing with respect to another and how to model the relationship between two variables. Here, the two variables are time and the level of CO_2.

Prerequisites

You should know how to graph data and fit a regression line to data on two variables.

Activity

There is concern that levels of CO_2 are rising in the atmosphere. In this activity, you will try to understand the rate of increase from 1958 to 1996 by using scatter plots. You will also fit straight lines to the complete data as well as subsets of the data and compare the slopes. You will use only the annual averages that are shown as the column labeled "CO_2 Level" in Table 1, or in the file **ModelsModelsModels.ftm** provided by your instructor.

Year	CO_2 Level	Year	CO_2 Level	Year	CO_2 Level
1958	315.174	1971	326.155	1984	344.247
1959	315.826	1972	327.293	1985	345.726
1960	316.748	1973	329.512	1986	346.975
1961	317.485	1974	330.079	1987	348.751
1962	318.297	1975	330.986	1988	351.313
1963	318.832	1976	331.986	1989	352.754
1964	319.038	1977	333.730	1990	354.037
1965	319.873	1978	335.336	1991	355.478
1966	321.210	1979	336.681	1992	356.292
1967	322.020	1980	338.515	1993	356.996
1968	322.890	1981	339.761	1994	358.880
1969	324.460	1982	340.959	1995	360.900
1970	325.517	1983	342.608	1996	362.574

Table 1: CO_2 levels measured at Mauna Loa, 1958–1996 (parts per million by volume [ppmv])
Source: http://cdiac.ornl.gov/ftp/ndp001/maunaloa.co2.

1. Plot the data and analyze the plots.
   a. Construct a scatter plot of the data using the year as the x-variable and the annual CO_2 level as the y-variable. What does the scatter plot of the data show? Are CO_2 levels steadily increasing?
   b. One way of studying the change in CO_2 is to calculate and then plot the year-to-year differences in CO_2. For example, the first point, the "difference" for 1959, would be (1959 CO_2 level) − (1958 CO_2 level).

      If the year-to-year differences were randomly scattered around one horizontal line, it would mean that the CO_2 levels were increasing at about the same rate throughout the entire period from 1958 to 1996. Calculate these differences and plot them by year. Now note the pattern of CO_2 levels from 1958 to 1972 and from about 1976 to 1996. Based on the plot, can we say that the patterns of changes in the annual CO_2 level are the same for the two periods?

Note: In Fathom, you can compute this difference by constructing a new attribute and giving it this formula:

CO_2Level − prev(CO_2Level)

2. Fit a model.
   a. Fit a straight line to the original data using the least-squares method and note the slope of the line. (In Fathom, select the graph from step 1a, then choose **Least-Squares Line** from the **Graph** menu.)
   b. One way to determine whether a straight line "fits" the data is to study the differences between the observed and fitted values. These differences are the *residuals*. Calculate the residuals for the model that uses all of the years. What does a positive residual signify? What does a negative residual signify?
   c. Plot the residuals in a scatter plot against the years. (In Fathom, select the graph, then choose **Make Residual Plot** from the **Graph** menu.) If the straight line is a good fit, there shouldn't be any visible pattern in the scatter plot and the residuals should be randomly scattered around zero. What does your residual plot show? Is it consistent with what you observed in step 1b?
   d. Since the CO_2 levels do not change the same way for the entire period, you will now try to describe the change with two straight lines. Fit a line using data from 1958 to 1972. Fit another line using data from 1976 to 1996. (In Fathom, make two different scatter plots. Select each graph in turn, choosing **Add Filter** from the **Data** menu. The formula for your filter might be, for example, **year < 1973.** Fathom calculates the least-squares line using only the points that "pass" the filter.)

3. Evaluate the fitted models.

   Look at the slopes of the fitted lines for the complete time period and for the two subsets. Complete the following sentences:

   From 1958 to 1996, the average annual CO_2 levels tended to ____________ at the rate of ____________ per year.

   From 1958 to 1972, the average annual CO_2 levels tended to ____________ at the rate of ____________ per year.

   From 1976 to 1996, the average annual CO_2 levels tended to ____________ at the rate of ____________ per year.

   Did the CO_2 levels increase at the same rate in the later years as in the earlier years?

4. Which of the three lines would give a better prediction of the CO_2 levels in 2002? Produce an estimate using the fitted line you think is best for prediction. What assumptions are you making when you determine the predicted value for 2002? If possible, go to the Web site provided for the data and read off the actual level for the year 2002.

Wrap-Up

There are different ways of deciding which function can model the relationship between two variables. What methods did you use? What did you learn about the steps to follow when fitting a model to data? The data in this activity indicate that annual levels of CO_2 are increasing. There are several reasons for this increase. However, scientists agree that one of the major culprits is the increase in fossil fuel consumption. The data in Table 2 give the gross energy consumption per capita in the United States. Gross energy includes the energy generated by primary fuels such as petroleum, natural gas, and coal; imports of their derivatives; and hydro and nuclear power.

Year	Gross Energy per Capita	Year	Gross Energy per Capita	Year	Gross Energy per Capita
1960	242.4	1971	326.9	1982	305.8
1961	242.1	1972	339.7	1983	301.3
1962	249.3	1973	350.5	1984	313.7
1963	255.3	1974	338.9	1985	309.6
1964	263.3	1975	326.4	1986	307.8
1965	271.2	1976	340.7	1987	311.8
1966	283.3	1977	346.5	1988	324.2
1967	289.9	1978	350.9	1989	325.3
1968	303.9	1979	349.7	1990	326.9
1969	316.7	1980	333.8	1991	334.7
1970	323.7	1981	322.7	1992	321.5

Table 2: Gross energy consumption in the United States, 1960–1992 (per capita, in millions of BTU per year)

1. Construct a scatter plot of the data. Would a straight line be a good model in this case?
2. Fit a straight line to the data and look at the residual plot. Does the residual plot confirm your answer to question 1?
3. Do you think that a nonlinear function might provide a better model? Explain why.

Extension

This activity was concerned with examining annual CO_2 levels to see how they changed over time. You fit a single straight-line model to the data and evaluated the fit by studying the residuals. The patterns in the residual plot and the plot of differences suggested that two lines would fit the data better than a single line. In many situations, it may not be feasible to fit one or more straight lines to the data; the "Wrap-Up" activity illustrated this point.

In this activity, we have limited ourselves to fitting straight lines to the CO_2 data. Could we have done better if we had fitted some other function? If you have the appropriate software, fit a quadratic or exponential function to the data and study the residual plot for this fit.

Mauna Loa

Background

PRINCIPAL INVESTIGATORS
Charles D. Keeling
Timothy P. Whorf
Scripps Institution of Oceanography
University of California
La Jolla, California 92093, U.S.A.

Air sample collection—Continuous. Air samples are collected from air intakes at the top of four 7-m towers and one 27-m tower.

Four samples are collected every hour.

Details are gicen in Keeling et al. (1982).

Measurement apparatus—Analyses of CO_2 concentrations ae made by using an Applied Physics Corporation nondispersive infrared gas analyzer with a water vapor freeze trap.

Data selection procedures—Data are selected for periods of steady hourly data to within ~0.5 ppmv; at least six consecutive hours of steady data were required to form a daily average.

Calibration gases used—CO_2-in-N_2 until December 1983 and CO_2-in-air from December 1983 to the present.

Scale of data reported—1987 WMO/Scripps mole fraction scale.

Data availability—These monthly and annual data, which are derived from daily "steady" data, are available from CDIAC. The monthly data through 1986, along with monthly data that have been adjusted to remove the seasonal effects are available in machine-readable form from CDIAC.

TREND

The Mauna Loa atmospheric CO_2 measurements constitute the longest continuous record of atmospheric CO_2 concentrations available in the world. The Mauna Loa site is considered one of the most favorable locations for measuring undisturbed air because possible local influences of vegetation or human activities on atmospheric CO_2 concentrations are condidered minimal and any influences from volcanic vents may be excluded from the records. The methods and equipment used to obtain these measurements have been essentially unchanged over the 31-year record.

Because of the favorable site location, continuous monitoring, and careful selection and scrutiny of the data, the Mauna Loa record is considered to be a precise record and a reliable indicator of the regional trend in the concentrations of atmospheric CO_2 in the middle layers of the troposphere. The Mauna Loa record shows a 12 percent increase in the mean annual concentration in 31 years, from 315 parts per million by volume of dry air (ppm) in 1958 to 352 in 1989 (Keeling et al. 1989).

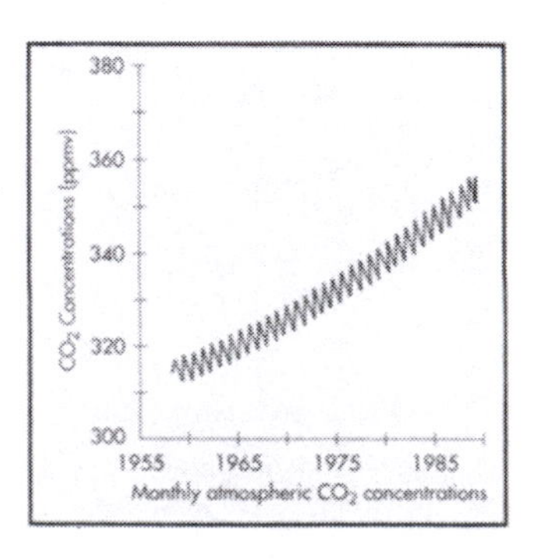

Monthly atmospheric CO_2 concentrations

Predictable Pairs

Relationships between categorical variables. Association in two-way tables.

Suppose you used a survey to gather information about students at your school, such as what proportion eat breakfast regularly, like 8 AM classes, have a job, or own a car. Would a "yes" response to a question about one of these characteristics help us predict what response that same person would give to another question? Is a person who eats breakfast likely to have an early class? Is a person who has a job likely to own a car? Looking for associations between categorical variables is an important part of analyzing frequency data. This lesson helps you see how to begin the process of measuring association.

Question

How can we tell if two categorical variables are associated?

Objectives

In this activity, you will begin to look at associations between two categorical variables by constructing two-way frequency tables. From these tables, you will calculate and interpret frequencies and relative frequencies of three types: joint, marginal, and conditional.

Prerequisites

You should be able to summarize categorical data using frequencies and relative frequencies for single variables. You should also know the difference between a proportion and a percent.

Activity

This is a class activity. You can collect all the data quickly from a few simple questions that will be asked of all class members.

1. Collecting the data

   As a class, formulate simple questions that can be answered either "yes" or "no." Some suggestions follow:

   Do you have a job outside of school?
   Do you own a motor vehicle?
   Do you participate regularly in a sport?
   Do you play a musical instrument?
   Did you see the recent movie (A) ___________?
   Did you see the recent movie (B) ___________?

2. Describing and interpreting the data

   a. Two questions with categorical responses are *associated* if knowledge of the response to one question helps predict the response to the other question. For example, knowing that a student owns an automobile may increase the chance that the student has a job. If that's the case, automobile ownership and job status are associated. From the list the class made, choose a pair of questions whose answers you think might be associated.

   b. Arrange the data for your pair of questions in a two-way frequency table, as illustrated in Table 1:

		Question A		
		Yes	No	Total
Question B	Yes	a	b	$a + b$
	No	c	d	$c + d$
	Total	$a + c$	$b + d$	$a + b + c + d$

Table 1

   From the class data you collected in step 1, which of the values in the two-way table can you fill in with the frequencies you already know?

   c. What additional information do you need to complete the table? Obtain this information and complete the table.

   d. Does it appear that the two questions you chose are associated? What features of the data in the two-way table give evidence for or against association?

3. Drawing conclusions

   a. What does a/n tell you? (Fill in your numbers for a and n.) What does c/n tell you? This is *joint information* of the two questions.

   b. What does $(a + c)/n$ tell you? This is *marginal information* on question A.

c. What does $(a + b)/n$ tell you? This is marginal information on question B.
d. What does $a/(a + c)$ tell you? This is *conditional information* on question B for those who are known to respond "yes" to question A.
e. What does $b/(b + d)$ tell you?
f. Which of the three types of information supplied by the two-way table (joint, marginal, or conditional) gives the greatest evidence for or against association between two questions in a pair? Explain your reasoning.

4. Summarizing results
Write a summary of your analysis for the data collected on two different pairs of questions used in your class. Address the issue of association in your summary.

Wrap-Up

1. Define a variable with ordered categorical responses, such as "How frequently do you drink coffee?" The responses could be categorized as never, seldom (less than once per week), or often. Then, define a second variable with similar response categories, such as "How often do you drink soft drinks?" Analyze the results of data collected from the class (or from another convenient group) by looking at a two-way table. A good way to set up such a table for ordered categories is presented in Table 2:

		Drink Coffee		
		Never	Seldom	Often
Drink Soft Drink	Often	a	b	c
	Seldom	d	e	f
	Never	g	h	i

Table 2

a. What type of association would be suggested by large values of (c, e, g)?
b. What type of association would be suggested by large values of (a, e, i)?
c. Do you see any relationship between the concepts of *association* between categorical variables and *correlation* between measurement variables? Explain.

2. The 1989 Teenage Attitudes and Practices Survey (see Allen and Moss, 1993) obtained completed questionnaires, either by telephone or mail, from a randomly selected group of 9965 12- to 18-year-olds living in households across the country. The teenagers answering the questions classified themselves as follows:
NS = never smoked
EX = experimented with smoking
FS = former smoker
CS = current smoker

One question was, "Do your peers care about keeping their weight down?" The data shown in Table 3 are the population projections (in thousands) calculated from the sample responses.

Peers Care about Keeping Weight Down	Respondents' Smoking: NS	EX	FS	CS
A Lot	6297	3613	197	2114
Somewhat	2882	1677	90	793
A Little	1441	625	16	354
Don't Care	1709	822	33	377

Table 3

a. Why would a question be formed in terms of peer behavior rather than as a direct question to the person being interviewed?

b. Approximately how many teenagers in the United States never smoked (as of 1989)?

c. What proportion of teenagers thought their peers cared a lot about keeping their weight down?

d. Approximately how many teenage current smokers were there in the United States in 1989? What proportion of them thought their peers cared a lot about keeping their weight down?

e. Among those that cared a lot about keeping their weight down, what proportion never smoked? What proportion were current smokers?

f. Do you think perceived peer attitudes toward keeping weight down are associated with the smoking status of the teenager? Calculate appropriate proportions to justify your answer.

3. Let's now apply our skills at analyzing frequency data to an important medical experiment. Joseph Lister, a British physician of the late 19th century, decided that something needed to be done about the high death rate from postoperative complications, which were mostly caused by infection. Based on the work of Pasteur, Lister thought that the infections had an organic cause. He decided to experiment with carbolic acid as a disinfectant for the operating room. Lister performed 75 amputations over a period of years. He used carbolic acid for 40 of the amputations, and no carbolic acid for the other 35. For the amputations done with carbolic acid, 34 of the patients lived; for those done without carbolic acid, 19 of the patients lived. Arrange these data on an appropriate table and discuss the association between mortality rate and the use of carbolic acid.

4. The ELISA (enzyme-linked immunosorbent assay) test for the presence of HIV antibodies in blood samples gives the correct result about 99.7% of the time for blood samples known to be HIV positive. The test is correct 98.5% of the time for blood samples known to be HIV negative.

a. Suppose 10% of 100,000 blood samples are HIV positive. All 100,000 samples are tested with ELISA. Set up a two-way table to show what the outcomes of the tests are expected to look like. What proportion of the samples that test positive do you expect to be positive in fact?

b. Repeat the preceding instructions for a set of 100,000 blood samples assumed to contain 20% that are HIV positive.

c. How do the results for steps 4a and 4b compare? Comment on the possible social implications of mass testing of blood samples.

Extensions

1. Is hand dominance associated with eye dominance? Here's how to determine your dominant eye: Hold your arms out in front of you and put your hands together so that there is just a small opening between them. Look through this opening at an object across the room. Now close one eye at a time. The eye with which you can still see the object is your dominant eye. You might want to try viewing several objects, one at a time, to help determine which eye is dominant. Now, collect data from a number of people on the following two-way table:

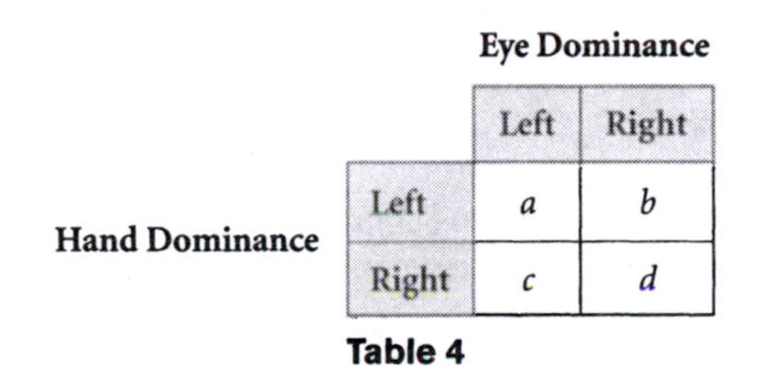

		Eye Dominance	
		Left	Right
Hand Dominance	Left	a	b
	Right	c	d

Table 4

a. Does there appear to be strong association between hand dominance and eye dominance?

b. What would large values for b and c tell you?

c. What would large values for a and d tell you?

2. Does aspirin really help prevent heart attacks? During the 1980s, approximately 22,000 physicians older than age 40 agreed to participate in a long-term health study for which one important question was to determine whether aspirin helps lower the rate of heart attacks (myocardial infarctions). The physicians were randomly assigned to receive either aspirin or a placebo as they entered the study. The method of assignment was equivalent to tossing a coin and sending the physician to the aspirin arm of the study if a head appeared on the coin. After the assignment, neither the participating physicians nor the medical personnel who treated them knew who was taking aspirin and who was taking placebo. This is called a *double-blind experiment.* (Why is the double blinding important in a study such as this?) The physicians were observed carefully for an extended period, and all heart attacks, as well as other problems that might occur, were recorded. All of the data can be summarized on two-way tables.

Other than aspirin, there are many variables that could affect the rate of heart attacks for the two groups of physicians. For example, the amount of exercise they get and whether they smoke are two prime examples of variables that should be controlled in the study so that the true effect of aspirin can be measured. Table 5 shows how the subjects were eventually divided according to exercise and to cigarette smoking.

		Aspirin	Placebo
Vigorous Exercise	Yes	7910	7861
	No	2997	3060
Cigarette Smoking	Never	5431	5488
	Past	4373	4301
	Current	1213	1225

Table 5
Source: "The final report on the aspirin component of the ongoing Physicians' Health Study," *N. Engl. J. Med.* (1989), 231(3):129–135.

a. Do you think the randomization scheme did a good job in controlling these variables? Explain in terms of association or lack of association.
b. Would you be concerned that the results for aspirin were unduly influenced by the fact that most of those taking aspirin were also nonsmokers? Explain.
c. Would you be concerned that the placebo group may have included too many subjects who did not exercise? Explain.

3. The results of this study report that 139 heart attacks developed among the aspirin users and 239 heart attacks developed in the placebo group. This was said to be a significant result in favor of aspirin as a possible preventive measure for heart attacks. To demonstrate this difference, place the data on heart attacks on an appropriate two-way table. (Remember, the 22,000 participants were about evenly split between the aspirin and placebo groups.) What are the appropriate conditional proportions to study if we want to compare the rates of heart attacks for the two treatment groups? Do these proportions turn out to be quite different?

An Example

Suppose you were asked for a show of hands on the question, "How many of you have jobs (outside of being a student)?" Then, you were asked for a show of hands on the question, "How many of you own a motor vehicle?" Setting up the appropriate two-way table and filling it in, you realized that the joint information was missing. You have to ask one more question of the class: "How many of you have jobs *and* own a motor vehicle?" With an answer to that one question, you can fill in the rest of the table.

Suppose the data from a group of 40 participants turned out as shown in Table 6:

		Job		
		Yes	No	Totals
Vehicle	Yes	8		10
	No			30
	Totals	26	14	40

Table 6

The marginal total of 10 tells us how many own vehicles, and the marginal relative frequency (or proportion) of 10/40 = 0.25 tells us that 25% of the group own vehicles. Similarly, 26/40 = 0.65 tells us that 65% of the class have jobs.

The joint frequency of 8 tells us that eight people have a job *and* own a vehicle. The joint relative frequency of 8/40 = 0.20 tells us that 20% of the whole group said "yes" to both of these questions.

Knowing the marginal totals and that one joint frequency, you can compute the rest. For example, you can tell that two people own a vehicle but have no job.

What can we say about vehicle ownership among those who have a job, as compared with those who do not have a job? We look at the conditional relative frequencies within each column. If we look only at people with a job in our group, the data show 8/26 = 0.31 as the proportion who own a vehicle. (This is often said to be the proportion who own vehicles *conditional* on the fact that they have jobs.) If we look at people who do not have jobs, then 2/14 = 0.14, or 14%, own vehicles.

We can calculate the conditional proportions across rows rather than down columns. Here, the proportion of people who have jobs conditional on the fact that they own a vehicle is 8/10 = 0.80. The proportion who have jobs conditional on the fact that they do not own a vehicle is 18/30 = 0.60.

So, after collecting and summarizing all of this data, can we answer the question, "Does having a job appear to be associated with owning a vehicle?" The marginal information sheds no light on the issue, because it describes the answers to the questions one at a time. The joint information tells us a little; 8 of 40 people saying "yes" to both questions may indicate that many who own vehicles may also have a job. But, what is *really* interesting is that these 8 come from only 10 vehicle owners in the entire group. Thus, the conditional proportion of 80% of vehicle owners having a job, as opposed to only 60% of nonvehicle owners having a job, shows that there appears to be an association between owning a vehicle and having a job.

We can analyze the conditional proportions the other way as well. The fact that 31% of those who have a job also own a vehicle, as opposed to only 14% of those who do not have a job, is, likewise, grounds for suggesting an association between the two variables.

What if the data turned out to look like those shown in Table 7?

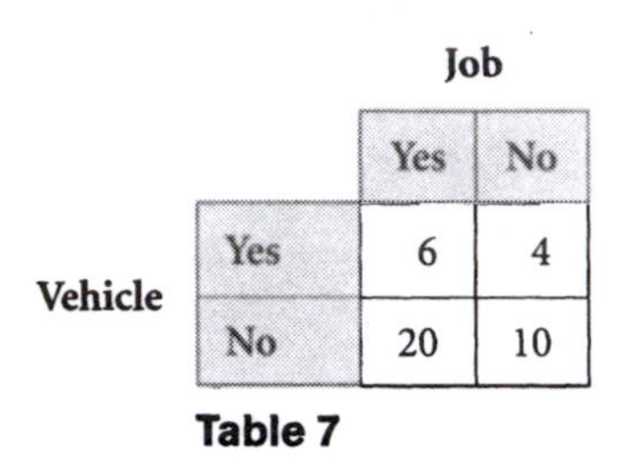

		Job: Yes	Job: No
Vehicle	Yes	6	4
	No	20	10

Table 7

Would this be a situation showing strong association between the two variables? Use arguments similar to those introduced previously to make the case for or against association.

Ribbon Charts

At any point in this activity, you might learn about the ribbon chart as a plot that shows these conditional frequencies in a visually compelling way. It is not important that you learn to make these plots by hand, but you should be able to interpret the graphics computer programs might make.

In a ribbon chart, the data are split along one categorical variable into rectangles whose areas are proportional to the frequencies. Each rectangle is then split, in the other direction, along the other categorical variable, again into rectangles with areas proportional to frequency.

You need to specify which variable comes first—it's the same as deciding whether to compute conditional proportions along the rows or along the columns.

The following ribbon charts show the data from the previous tables.

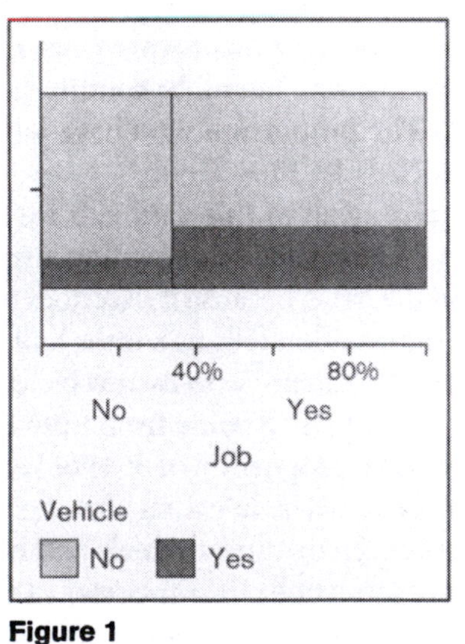

Figure 1

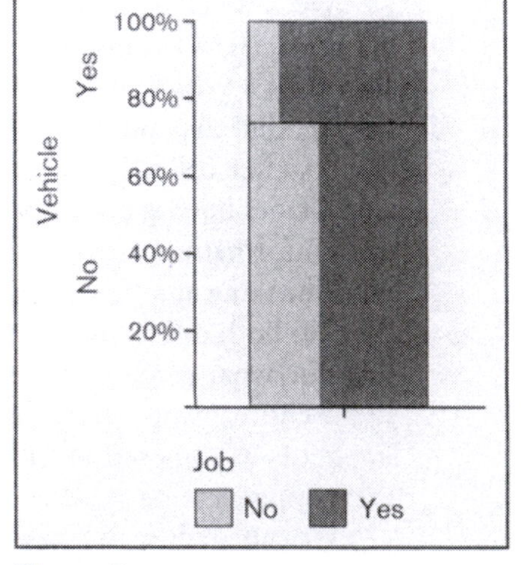

Figure 2

You can see how the chart helps compare proportions between groups of different sizes and how the various rectangles have areas proportional to the relevant frequencies. What may not be as obvious is that when there is no association, the breaks cutting across the rectangles will form a single, unbroken line; in this case, the four rectangles in each chart would meet at a point.

Assessment Questions

1. Find an article from a newspaper or magazine that contains responses to categorical variables. (Most opinion polls are set up this way.) Discuss possible associations among the variables in the article. Can these associations be measured from the data presented in the article? (Often they cannot because the articles tend not to give the joint information.)

2. Another question on the survey of teenagers was, "Do your peers care about staying away from marijuana?" The population-projected frequencies (in thousands) are given in Table 8.

Peers Care about Marijuana Abstention	Respondents' Smoking: NS	EX	FS	CS
A Lot	7213	2693	75	857
Somewhat	2482	1861	109	1102
A Little	744	542	27	298
Don't Care	1878	1550	119	1312

Table 8

   a. What proportion of those who smoked think their peers care a lot about staying away from marijuana?
   b. What proportion of the current smokers think their peers care a lot about staying away from marijuana?
   c. Among those who think their peers care a lot about staying away from marijuana, what proportion have never smoked?
   d. Among those who think their peers don't care about staying away from marijuana, what proportion have never smoked?
   e. Do you think perceived peer attitudes toward staying away from marijuana are associated with the smoking status of the teenager? What proportions help justify your answers?

3. Additional questions on the Teenage Attitudes and Practices Survey were asked directly of those being interviewed. Some of the data are reported in the form of percentages, rather than frequencies. Two examples follow (Tables 9 and 10):

	Respondents' Smoking			
	NS	EX	FS	CS
Yes	12.0	18.7	29.8	46.5
No	84.9	78.5	68.9	51.7
Don't Know	3.0	2.5	1.6	1.6

Table 9: "Do you believe that cigarette smoking helps reduce stress?"

	Respondents' Smoking			
	NS	EX	FS	CS
Yes	80.1	78.8	80.1	80.5
No	17.3	18.8	17.3	16.7
Don't Know	2.5	2.3	2.6	2.6

Table 10: "Do you believe almost all doctors are strongly against smoking?"

a. How were these percentages calculated? What do they mean? Are these percentages joint, marginal, or conditional?

b. Do the opinions on whether or not cigarette smoking helps reduce stress appear to be associated with the smoking status of the person responding? Write a paragraph justifying your answer.

c. Do the opinions on doctors being against smoking appear to be associated with the smoking status of the person responding? Write a paragraph justifying your answer.

4. Heart attacks aren't the only cause for concern in the Physician's Health Study. Another is that too much aspirin can cause an increase in strokes. Among the aspirin users in the study, 119 had strokes during the observation period. Within the placebo group, only 98 had strokes. Place these data on an appropriate two-way table and comment on the association between aspirin use and strokes, as compared with the association between aspirin use and heart attacks.

5. What about smoking? How does it relate to heart attacks and the use of aspirin? Tables 11 and 12 show the number of heart attacks for each treatment group separated out according to whether the participant was a current smoker or had never smoked.

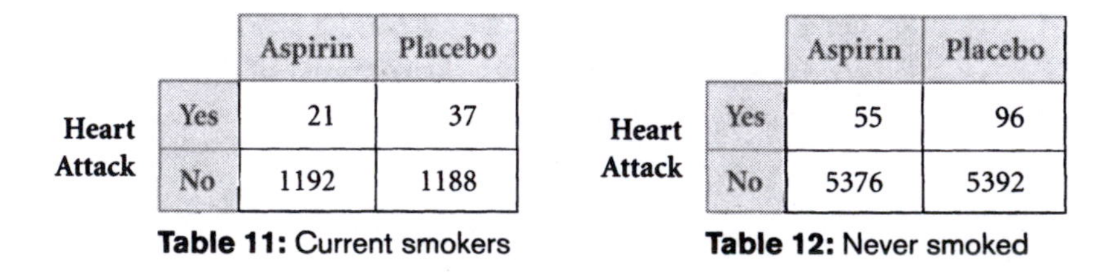

		Aspirin	Placebo
Heart Attack	Yes	21	37
	No	1192	1188

Table 11: Current smokers

		Aspirin	Placebo
Heart Attack	Yes	55	96
	No	5376	5392

Table 12: Never smoked

Is aspirin as effective at preventing heart attacks among current smokers as it is among those who never smoked? What can we say about the rate of heart attacks among the current smokers compared with that of those who never smoked?

Ratings and Ranks

The relationship between ratings and ranks.

"An outfit calling itself 'Morgan Quitno Press' recently ranked the 50 United States in order of intelligence, and I am TICKED OFF. My state, Florida, came in 47th. Can you believe that? Forty-seventh! How dare they? How dare they suggest that Florida is more intelligent than three other states? No way!"

—Dave Barry in *The Miami Herald,* October 20, 2002

Ratings and ranks—we love them and regularly use them to gauge our thoughts and feelings about places and events in our lives. Automobiles are rated on various performance criteria and then ranked against one another. Athletes are similarly rated on their performance and then ranked in comparison to others. In Olympic figure skating, for example, a high rating produces a rank close to the top of the scale. Let's take an example. The Nielsen Co. reports that a certain TV show has 23% of the market for its time slot. This information is in the form of a rating. The Nielsen Co. reports that the TV show has the second highest percentage of the viewing audience for that time slot. Now we have a rank.

In the case of Morgan Quitno Press, how did they rank the states? For the "Smartest State Award" (which Connecticut won), Morgan Quitno Press looked at statistics that rate the states on many different criteria, combined them into a single rating, and then ranked the states using these combined ratings. This is such a common process—and so treacherous—that it deserves our attention.

Question

What are ratings and what are ranks? How are the two related?

Objectives

In this activity, you will see how people combine various variables to form a numerical rating of a person, place, or object. The way they combine these variables is often quite arbitrary, and you should take this into account when you interpret ratings. You will discover that changes in the relative weights of variables making up a rating can have a dramatic effect on the conclusions. Finally, you will see that it is convenient to turn ratings into ranks when you need to summarize information for potential users.

Prerequisites

You need to know about weighted averages, although they could be explained in the context of this activity.

Activity

1. Understanding ratings

   Suppose we wanted to make our own "Smartest States" report. First, it's not obvious what we mean by "smartest." The best student test scores? The most accomplished adults? Perhaps "smart" isn't about schools at all. We could use many variables to determine which state is smartest, but we will use only five (and we will stick to education):

   Percent bachelors (PB): the percentage of persons 25 or older who hold bachelor's degrees (1990 U.S. Census)

   Scholastic Aptitude Test scores (SAT): the mean SAT score of high-school graduates in 1999

   Graduation rate (GR): the percentage of entering high-school freshmen who graduate in 4 years, for the class of 2000

   Student/teacher ratio (STR): the student/teacher ratio in public schools for the 2000–2001 school year

   Bloat (BL): our favorite—the number of school administrators in the state divided by the number of school librarians (Some analysts suggest it as one measure of a state's commitment to services for students.)

   a. For which variables are high ratings good and low ratings bad?
   b. Which variables might tend to produce higher ratings in more rural states?
   c. Which would have the greater effect on the "bloat" rating: an increase of 1000 librarians in a state or an increase of 1000 administrators? Explain.
   d. States vary widely in the proportion of students who take the SAT. What effect (if any) does this fact have on using SAT as a variable?
   e. Most actual class sizes are larger than the student/teacher ratio. Why is that?

2. Determining ranks from ratings

   Pick eight states: Connecticut, Arkansas, Florida, California, and four more (including your own) and copy the data from Table 1 onto a separate sheet or into Table 2.

State	PB	SAT	GR	STR	BL
Alabama	15.7	1116	81.6	15.4	6
Alaska	23	1030	93.3	16.9	17.4
Arizona	20.3	1049	73.5	19.8	11.4
Arkansas	13.3	1119	84.1	14.1	4.4
California	23.4	1011	82.5	20.6	48.1
Colorado	27	1076	81.6	17.3	11.7
Connecticut	27.2	1019	91.7	13.7	10.7
Delaware	21.4	1000	91	15.3	11.2
District of Columbia	33.3	972	88	13.9	10.6
Florida	18.3	997	84.6	18.4	13.6
Georgia	19.3	969	83.5	15.9	7.5
Hawaii	22.9	995	91.8	16.9	5.3
Idaho	17.7	1082	86.4	17.9	11.5
Illinois	21	1154	87.1	16.1	13.2
Indiana	15.6	994	89.4	16.7	10.8
Iowa	16.9	1192	90.8	14.3	10.9
Kansas	21.1	1154	90.4	14.4	6.2
Kentucky	13.6	1094	86.2	16.8	6.4
Louisiana	16.1	1119	82.1	14.9	5.6
Maine	18.8	1010	94.5	12.5	13
Maryland	26.5	1014	87.4	16.3	7.9
Massachusetts	27.2	1022	90.9	14.5	13.7
Michigan	17.4	1122	89.2	18	10.8
Minnesota	21.8	1184	91.9	16	10.4
Mississippi	14.7	1111	82.3	16.1	6.5
Missouri	17.8	1144	92.6	14.1	10.1
Montana	19.8	1091	91.1	14.9	5.2
Nebraska	18.9	1139	91.3	13.6	6.1
Nevada	15.3	1029	77.9	18.6	10.1
New Hampshire	24.4	1038	85.1	14.5	7.8
New Jersey	24.9	1008	90.1	13.1	12.4
New Mexico	20.4	1091	83	15.2	21.7
New York	23.1	997	86.3	13.9	13.5
North Carolina	17.4	986	86.1	15.5	2.7
North Dakota	18.1	1199	94.4	13.4	7
Ohio	17	1072	87.7	15.5	20.9
Oklahoma	17.8	1127	85.7	15.1	7.7
Oregon	20.6	1050	82.3	19.4	12.5
Pennsylvania	17.9	993	89	15.5	9.4
Rhode Island	21.3	1003	87.9	14.8	33.7
South Carolina	16.6	954	85.1	14.9	7.7
South Dakota	17.2	1173	92	13.7	10
Tennessee	16	1112	89	14.9	8.2
Texas	20.3	993	79.4	14.8	8.8
Utah	22.3	1138	90	21.9	12.2
Vermont	24.3	1020	90.8	12.1	6.3
Virginia	24.5	1007	87.3	12.5	8.8
Washington	22.9	1051	87.4	19.7	7.9
West Virginia	12.3	1039	89.6	13.7	9.4
Wisconsin	17.7	1179	90	14.1	7.2
Wyoming	18.8	1097	86.5	13.3	12

Table 1: State education data

State	Weighted Average	PB	PB Rank	SAT	SAT Rank	GR	GR Rank	STR	STR Rank	BL	BL Rank
Arkansas		13.3		1119		84.1		14.1		4.4	
California		23.4		1011		82.5		20.6		48.1	
Connecticut		27.2		1019		91.7		13.7		10.7	
Florida		18.3		997		84.6		18.4		13.6	

Table 2

a. Review your answer to step 1b. Do the data suggest that you were correct?

b. For which variable are the ratings most variable? For which are the ratings least variable? Does this seem reasonable? Explain.

c. With 1 denoting most desirable and 8 denoting least desirable, rank the states with regard to percentage of bachelors degrees (PB). Enter the ranks in the **PB rank** column. What features of the data are lost when you go from the actual rating to the rank?

d. Rank the states within each of the other four variables, with 1 denoting most desirable. Write the ranks on the table next to the ratings in the appropriate columns.

e. Discuss the differences and similarities in rankings you made in step 2d. What are some advantages to reporting the ranks rather than the ratings?

3. Combining ranks

You will now be making computations with the data from your table. The most efficient way to accomplish this is to enter the data into a graphing calculator as lists or into a computer spreadsheet. The goal of this investigation is for you to come up with your personal overall ranking of the eight states to determine which is the "smartest." The first step is to establish the relative importance of the five variables.

a. Gather 10 pennies. Divide the pennies among the five variables on Table 2 according to the relative importance you assign to the variables. (Place pennies on the data table.) For example, if you think that all five variables are equally important, then lay two pennies on each column. If you think that SATs and bloat are equally important and nothing else matters, place five pennies on SATs and five on bloat. The number of pennies assigned to a variable divided by 10 (which makes a convenient decimal) is your personal weighting for that variable.

b. Using your weights as established in step 3a, calculate the weighted average of the ranks for each state. Enter those averages into the appropriate column.
c. Rank the states according to your weighted average ranks from step 3b.
d. Compare your ranking of the states with those of other class members. Is there any one state that always ranks first?

Wrap-Up

1. Discuss the difference between a rating and a rank.
2. What other variables might be important in rating how "smart" a state is?
3. What questions might you ask when someone reports a new study ranking desirable spots for a vacation or desirable occupations?
4. Table 3 shows the NFL quarterbacks with at least 300 passing attempts during the 2001 regular season. In the table, they're listed in order of number of attempts; but the highest-rated passer that year was Kurt Warner, with a rating of 101.4. The formula used to produce these ratings is

$$R = (1/24)\{50 + 20C + 80T - 100I + 100(Y/A)\}$$

where

R = the rating

C = percentage of completions

T = percentage of touchdowns

I = percentage of interceptions

Y = yards gained by passing

A = passing attempts

a. Suppose you were a quarterback who played only one play in the entire season. You came in, threw one pass, which was complete for a 15-yard gain. What would your rating be?
b. Using a spreadsheet on a computer or a graphing calculator, add columns to the data table for C, T, I, and Y/A. (If you're using Fathom, the data are provided by your instructor and are in the file **2001 QB ratings.ftm.**)
c. Using the constructed data columns, find the ratings (R) for the other quarterbacks on the list. Who is the second-highest-rated passer?
d. Note that the list is for the 2001 season only. Find the passing statistics for your favorite NFL quarterback from last season. How does he compare with these quarterbacks?

Player	Team	Games Played	Passing Attempts	Passing Completions	Yards Gained	Touchdowns	Interceptions	Sacks
Jon Kitna	Cincinnati	16	581	313	3216	12	22	25
Kerry Collins	New York Giants	16	568	327	3764	19	16	36
Brad Johnson	Tampa Bay	16	559	340	3406	13	11	44
Aaron Brooks	New Orleans	16	558	312	3832	26	22	50
Rich Gannon	Oakland	16	549	361	3828	27	9	27
Peyton Manning	Indianapolis	16	547	343	4131	26	23	29
Kurt Warner	St. Louis	16	546	375	4830	36	22	38
Chris Weinke	Carolina	15	540	293	2931	11	19	26
Jake Plummer	Arizona	16	525	304	3653	18	14	29
Trent Green	Kansas City	16	523	296	3783	17	24	39
Doug Flutie	San Diego	16	521	294	3464	15	18	25
Brett Favre	Green Bay	16	510	314	3921	32	15	22
Jeff Garcia	San Francisco	16	504	316	3538	32	12	26
Donovan McNabb	Philadelphia	16	493	285	3233	25	12	39
Mark Brunell	Jacksonville	15	473	289	3309	19	13	57
Elvis Grbac	Baltimore	14	467	265	3033	15	18	28
Tim Couch	Cleveland	16	454	272	3040	17	21	51
Brian Griese	Denver	15	451	275	2827	23	19	38
Jay Fiedler	Miami	16	450	273	3290	20	19	27
Kordell Stewart	Pittsburgh	16	442	266	3109	14	11	29
Vinny Testaverde	New York Jets	16	441	260	2752	15	14	18
Steve McNair	Tennessee	15	431	264	3350	21	12	37
Tom Brady	New England	15	413	264	2843	18	12	41
Jim Miller	Chicago	14	395	228	2299	13	10	11
Daunte Culpepper	Minnesota	11	366	235	2612	14	13	33
Chris Chandler	Atlanta	14	365	223	2847	16	14	41
Charlie Batch	Detroit	10	341	198	2392	12	12	33
Matt Hasselbeck	Seattle	13	321	176	2023	7	8	38
Alex Van Pelt	Buffalo	12	307	178	2056	12	11	14

Table 3: NFL passing data, 2001

e. Discuss the relative weights in the formula for quarterback ratings. You might begin by answering the following questions:

   i. Why does one variable have a minus sign in front of it?

   ii. How much more important is a touchdown than a simple completion?

   iii. Does a touchdown carry more weight than an interception?

   iv. Why use yards per attempt rather than just total yards passing?

   v. If you were to include number of sacks in the rating, how might you do it? Could that change who was rated best?

Extensions

1. Discuss ways we could get a composite ranking of colleges. Discuss their advantages and disadvantages.

2. Find a copy of the *Places Rated Almanac.*
   a. What method do they use to produce a composite ranking of the cities in the United States?
   b. Select cities of interest to you and repeat the second investigation ("Determining ranks from ratings") as described in the second activity. You may want to use different variables, according to your interests.

Assessment Questions

1. The NCAA uses the following formula to establish a rating for college quarterbacks:

$$\text{Rating} = (\%\text{COMP}) + 3.3(\%\text{TD}) - 2.0(\%\text{INT}) + 8.4(\text{YDS/ATT})$$

   a. Compare this formula with the one the NFL uses. Comment on the similarities and differences between the two. Does an interception carry more weight in the NFL or in the NCAA?
   b. Find the statistics for two college quarterbacks and produce their ratings. How do they rank compared with each other?

2. Select five brands of automobiles that are of interest to you.
   a. Obtain data for each on price, resale value, gas mileage, interior room, and trunk space.
   b. Weight these five variables according to their importance to you.
   c. Calculate a composite rank for each automobile and rank them on this composite measure.
   d. Find another student who has a different ranking and comment on why the two of you differ.

II

~

Planning a Study

Random Rectangles

Sampling bias. The importance of random sampling.

Results from polls and other statistical studies reported in the media often emphasize that the samples were randomly selected. Why the emphasis on randomization? Couldn't a good investigator do better by carefully choosing respondents to a poll so that various interest groups were represented? Perhaps, but samples selected without objective randomization tend to favor one part of the population over another. For example, polls conducted by sports writers tend to favor the opinions of sports fans. This leaning toward one side of an issue is called *sampling bias.* In the long run, random samples seem to do a good job of producing samples that fairly represent the population. In other words, randomization *reduces* sampling bias.

Question

How do random samples compare with subjective samples in terms of sampling bias?

Objectives

In this activity, you will compare subjective (or judgmental) samples with random samples and see which better represent the population. The goal is to learn why randomization is an important part of data collection.

Prerequisites

You should know how to use random numbers to select random samples. In addition, you should understand the difference between a population and a sample and know why sampling is essential in most statistics problems.

Activity

1. Judgmental samples
   a. Find the accompanying sheet of rectangles in your book (page 84) but keep this sheet covered until the instructor gives the signal to begin. When signaled to do so, look at the sheet for a few seconds and estimate the average area of the rectangles on the sheet. (The unit of measure is the background square. Thus, rectangle 33 has area $4 \times 3 = 12$.) Write down your estimate.
   b. Select five rectangles that, in your judgment, are representative of the rectangles on the page. Write down the area for each of the five. Compute the average of the five areas and compare it with your estimate. Are the two numbers close?
   c. The instructor will provide you with all class members' estimates and the averages of their subjective samples of five rectangles. Display the two sets of data on separate dot plots. Comment on the shapes of these distributions and where they center. Why is the center an important point to consider?

2. Random samples
   a. Use a random number table (or a random number generator in a computer or calculator) to select five distinct random numbers between 00 and 99. Find the five rectangles with these numbers and mark them on the sheet. This is your random sample of five rectangles.
   b. Compute the areas of these five sampled rectangles and find the average. How does the average of the random sample compare with your estimate in step 1a? How does it compare with your average for the subjective sample?
   c. The instructor will collect the averages from the random samples of size 5 and construct a dot plot. How does this plot compare with the plots of the estimated values and the averages from the subjective samples in terms of center? In terms of spread?

3. Data analysis
   a. From the data the instructor has provided for the whole class, calculate the mean of the sample averages for the subjective samples and for the random samples. How do these centers of the distributions of means compare?
   b. Calculate the standard deviation of the averages for the subjective samples and for the random samples. How do the spreads of the distributions of means compare?
   c. Having studied two types of sampling, subjective and random, which method do you think is doing the better job? Why?

4. Sampling bias
   Your instructor will now place the true average area of the rectangles on each of the plots.
   a. Do either of the plots have a center that is very close to the true average?

b. Do either of the plots have a center that is larger than the true average?

c. Discuss the concept of bias in sampling and how it relates to the two sampling methods, subjective and random, you just used.

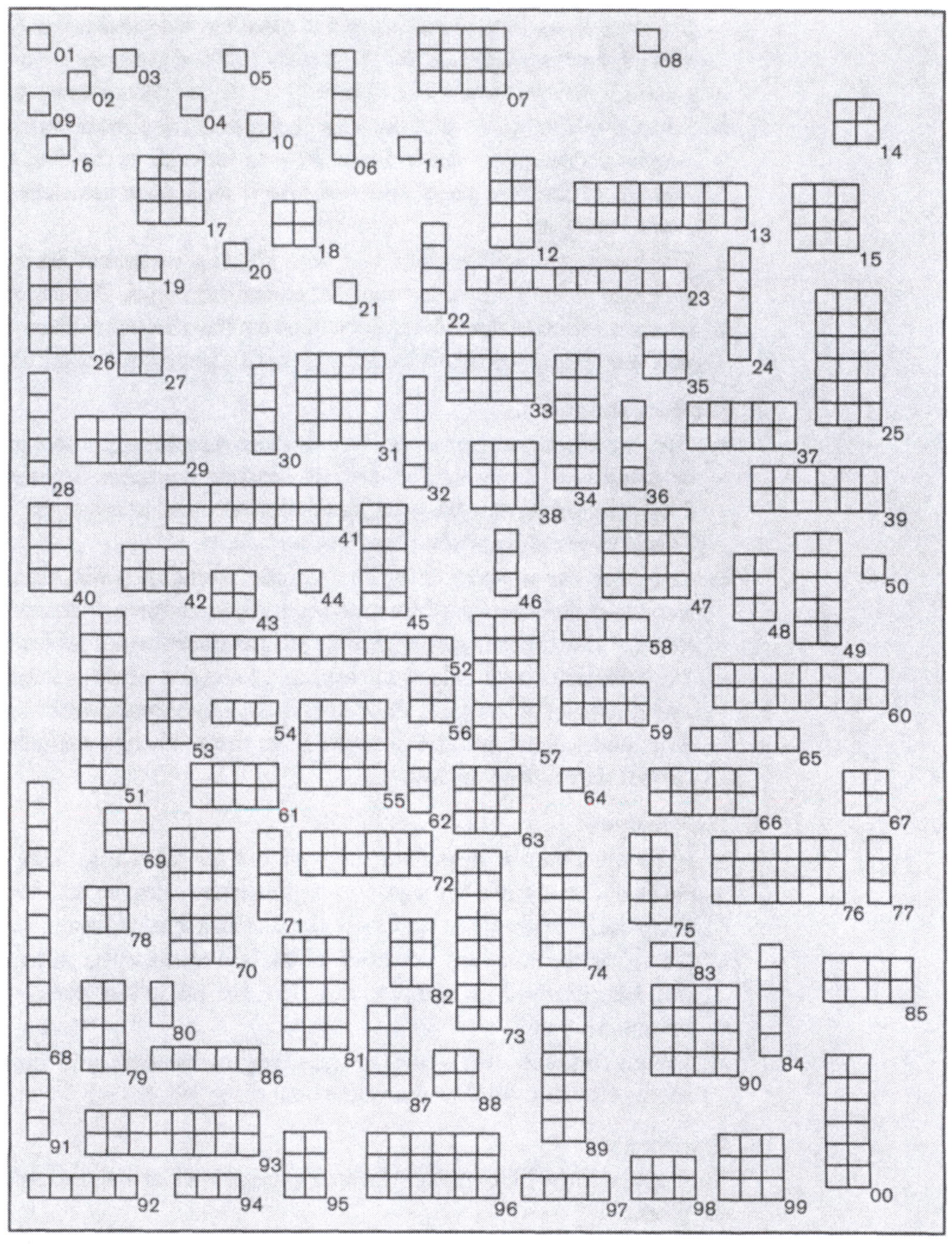

Figure 1

39634	62349	74088	65564	16379	19713	39153	69459	17986	24537
14595	35050	40469	27478	44526	67331	93365	54526	22356	93208
30734	71571	83722	79712	25775	65178	07763	82928	31131	30196
64628	89126	91254	24090	25752	03091	39411	73146	06089	15630
42831	95113	43511	42082	15140	34733	68076	18292	69486	80468
80583	70361	41047	26792	78466	03395	17635	09697	82447	31405
00209	90404	99457	72570	42194	49043	24330	14939	09865	45906
05409	20830	01911	60767	55248	79253	12317	84120	77772	50103
95836	22530	91785	80210	34361	52228	33869	94332	83868	61672
65358	70469	87149	89509	72176	18103	55169	79954	72002	20582
72249	04037	36192	40211	14918	53437	60571	40995	55006	10694
41692	40581	93050	48734	34652	41577	04631	49184	39295	81776
61885	50796	96822	82002	07973	52925	75467	86013	98072	91942
48917	48129	48624	48248	91465	54898	61220	18721	67387	66575
88378	84299	12193	03785	49314	39761	99132	28775	45276	91816
77800	25734	09801	92087	02955	12872	89848	48579	06028	13827
24028	03405	01178	06316	81916	40170	53665	87202	88638	47121
86558	84750	43994	01760	96205	27937	45416	71964	52261	30781
78545	49201	05329	14182	10971	90472	44682	39304	19819	55799
14969	64623	82780	35686	30941	14622	04126	25498	95452	63937
58697	31973	06303	94202	62287	56164	79157	98375	24558	99241
38449	46438	91579	01907	72146	05764	22400	94490	49833	09258
62134	87244	73348	80114	78490	64735	31010	66975	28652	36166
72749	13347	65030	26128	49067	27904	49953	74674	94617	13317
81638	36566	42709	33717	59943	12027	46547	61303	46699	76243
46574	79670	10342	89543	75030	23428	29541	32501	89422	87474
11873	57196	32209	67663	07990	12288	59245	83638	23642	61715
13862	72778	09949	23096	01791	19472	14634	31690	36602	62943
08312	27886	82321	28666	72998	22514	51054	22940	31842	54245
11071	44430	94664	91294	35163	05494	32882	23904	41340	61185
82509	11842	86963	50307	07510	32545	90717	46856	86079	13769
07426	67341	80314	58910	93948	85738	69444	09370	58194	28207
57696	25592	91221	95386	15857	84645	89659	80535	93233	82798
08074	89810	48521	90740	02687	83117	74920	25954	99629	78978
20128	53721	01518	40699	20849	04710	38989	91322	56057	58573
00190	27157	83208	79446	92987	61357	38752	55424	94518	45205
23798	55425	32454	34611	39605	39981	74691	40836	30812	38563
85306	57995	68222	39055	43890	36956	84861	63624	04961	55439
99719	36036	74274	53901	34643	06157	89500	57514	93977	42403
95970	81452	48873	00784	58347	40269	11880	43395	28249	38743
56651	91460	92462	98566	72062	18556	55052	47614	80044	60015
71499	80220	35750	67337	47556	55272	55249	79100	34014	17037
66660	78443	47545	70736	65419	77489	70831	73237	14970	23129
35483	84563	79956	88618	54619	24853	59783	47537	88822	47227
09262	25041	57862	19203	86103	02800	23198	70639	43757	52064

Table 1: Random number table

Wrap-Up

1. Discuss the difference between sampling bias and measurement bias. Give examples of statistical studies in which each is an important consideration. Are either of these biases reduced appreciably by increasing the sample size?

2. Find a printed article that reflects sampling bias. (The article that follows on page 87 is an example.) How could the sampling bias have affected the conclusions reported in the article?

Extensions

1. Using the same sheet of rectangles used earlier, select multiple random samples of 10 rectangles each and compute the average area for each sample. Plot these averages and compare the plot to the one for the random samples of size 5 with regard to the following:
   - **a.** Center
   - **b.** Spread
   - **c.** Shape

2. Obtain a map of your state that shows the county (or parish) boundaries. Your job is to show the geography department how to select a random sample of counties for purposes of studying land use. How would you select the sample? What might cause bias in the sampling process?

Assessment Questions

1. You are to design a sampling plan to measure one of the following for the community in which you live:
   - **a.** Average number of TV sets per household
   - **b.** Average number of pets per student
   - **c.** Percentage of foreign autos driven by students at your school

   Choose one of the items and discuss how you would obtain the data. Mention the role of randomization and the possibility of bias in your data collection plan.

2. Critique the following article with respect to the sampling issues involved and the conclusions reached.

The Popular Science Nuclear Power Poll

Landslide for Safer Nuclear

By a stunning majority, voters in the POPULAR SCIENCE poll [Cast Your Vote On Next Generation Nuclear Power, April '90] favor building a new generation of nuclear power plants. Responding to our detailed report on nuclear plant designs and concepts claimed to be safer than existing nuclear generating stations, voters from 38 states took part in the poll and more than 5,600 persons cast electronically recorded ballots on the main issue: Should the United States build more fossil-fuel power plants or the new so-called safe nuclear generators to meet the energy crisis of the '90s? The pro-nuke voters prevailed by a margin of more than six to one.

Why did the vote turn out in this fashion? In the follow-up interviews, voters said that nuclear power in a new package could meet energy needs in a safe, environmentally friendly way. Said one voter in Oklahoma City: "It's also better than acid rain." The Oklahoma reader also said he was in favor of building "little ones rather than mega ones." A New Jersey minister told us that "more people have gotten sick from Jane Fonda movies than from nuclear power."

A huge majority of voters also indicated a preference for nuclear power research over alternate energy. In addition, the tally showed disapproval of the government's energy policies.

The vote was clear-cut despite an attempt on the part of Westinghouse, one of the world's major nuclear power plant designer, to stuff the electronic ballot box. Westinghouse set up a special speed dialing number, enabling its Pittsburgh headquarters employees to call our 900 poll number (which cost voters 75 cents a minute) for free, and then publicized the article and poll in an internal newsletter. Westinghouse employees responded with a total of 20 percent of the total votes cast in our poll, according to our research.

(POPULAR SCIENCE protested the action of Westinghouse, criticizing the nuke manufacturer for a sinister and cynical disregard for an open and honest forum of free ideas. Richard Slember, vice president of Westinghouse's energy systems, replied: "We in no way tried to influence any votes, but we did want our employees to know about the poll and to have the opportunity to participate.")

Even when these votes are discarded, the result is a landslide favoring the building of a new generation of safer nuclear plants. If we assume that all the votes originating from Westinghouse were pro nukes and disqualify them, the margin is still about five to one in favor of the new nuclear plants.

Voting on the second question, which posed the issue of safety, and on the question that asked if respondents would rather put up with blackouts than see new nukes built, revealed that there is less concern about safety than might be supposed. And there's definitely no strong sentiment for cutting back on energy consumption.

A California student explained his pro vote this way: "You're more likely to be struck by lightning twice than hurt by nuclear power." And a Coloradan simply explained his pro vote with: "It's a necessary evil, we have to have it."

Our April cover article also provoked the greatest volume of letters we've received for one story in recent memory. In these, there was a dis-

(continued)

tinct trend *against* the proposed new generation of nuclear power plants, although the ratio was less pronounced: 60 percent to 40 percent (a sampling appears in the box at right).

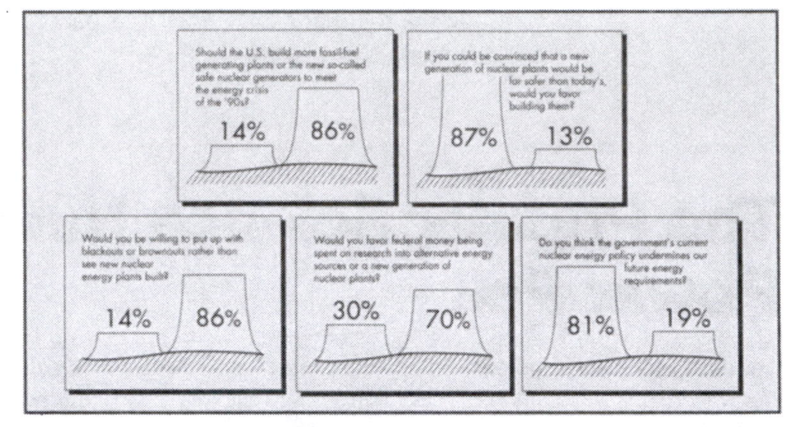

Voters who opposed the new nukes also had strong feelings on the subject. One negative vote was cast by a Reno, Nev. cook who declared, "Scientists are destroying our society. I've lost all confidence in these people."

Some voters simply told us that there are no easy solutions to the energy problem—that fossil and nuclear fuels both posed threats to mankind. But virtually all of these comments came from people who voted in favor of the new nukes.

The voting also revealed a preference for spending more money on nuclear-energy research rather than on alternate-energy research. However, in our exit poll calls, there were a number of respondents who indicated that alternate energy is a viable alternative to nuclear and fossil fuels. And, by a margin of four to one, the voters also expressed a deep distrust of the government's energy policies. The vote indicates that there is a strong feeling that the government's energy policy undermines future needs.

Copies of the April article, the results of our poll, and the letters will be forwarded to leading policy makers, including President Bush.

Source: Popular Science, August 1990.

The Rating Game

Analyzing ratings that are results from a questionnaire you design.

As consumers, we are often faced with making decisions based on ratings, whether they are ratings of movies, the Nielsen Co. ratings of TV shows, or the car ratings in *Consumer Reports. U.S. News and World Report* annually rates colleges and universities. To establish ratings, the magazine uses data such as average SAT/ACT scores of freshmen, the acceptance rate, financial resources, and the ratings of schools by the presidents, deans, and admissions directors. The elite schools compete with each other for the top ratings and use them to attract the best students. Do you know how your school did in this rating scheme?

Of course, an overall rating from a magazine will not reflect your day-to-day experiences at the school. This activity asks you to determine your fellow students' level of satisfaction with three different aspects of your school that are important to you: academic programs, physical plant, and extracurricular activities.

Question

How satisfied is the student body with the academic programs, physical plant, and extracurricular activities?

Objectives

In this activity, you will gain some experience in designing questionnaires and summarizing and analyzing score data.

Prerequisites

You need to know how to construct data displays, such as dot plots, stem and leaf diagrams, and box plots. You need to be able to summarize data using means and medians and using the range and the standard deviation.

Activity

1. Work in groups to design a questionnaire that addresses these three categories: academic programs, physical plant, and extracurricular activities. In each category, include at least three questions asking students to give a score between 0 and 100, where 0 indicates the lowest level of satisfaction and 100 the highest level of satisfaction. This means that you have to word the questions so that the response is a score and not a "yes" or "no." For example, if you want to find out how students feel about the scheduling of the courses, a sample question is, "How satisfied are you with the times at which your courses are scheduled?" Following are examples of issues you could consider in each category:

   Academic programs
   Quality of instruction
   Quality and variety of course offerings
   Availability of required courses
   Scheduling of the courses
   Flexibility in designing your program

   Physical plant
   Condition of classrooms
   Maintenance of dormitories
   Access to libraries/gym
   Parking

   Extracurricular activities
   Recreational/cultural programs on campus
   Recreational/cultural programs off campus
   Student clubs and organizations

2. After you have designed the questionnaire, do a pilot survey. Ask some of your friends to respond to the questionnaire. Change the wording of questions as needed if their responses do not give you the information you want.

3. Administer your questionnaire to the rest of the class or to a randomly chosen sample of students outside the class.

4. Analyze the responses using graphs such as dot plots, stem and leaf diagrams, and box plots. Summarize the responses with means, medians, ranges, and standard deviations. Combine the summaries within each category to derive an overall rating for the category. Be prepared to justify your method of combining the response summaries.

5. Write a report that includes a discussion of your questions, the method you used to analyze the responses, and your conclusions.

Wrap-Up

In any questionnaire, the wording of the questions, the scale used for the responses, the length of the questionnaire, and the order in which the questions are asked can all influence the responses.

1. Select a question from your questionnaire that resulted in responses within the middle of the rating scale and change the wording to draw mostly negative or mostly positive responses.

2. This activity asks you to use a rating scale from 0 to 100. From observing your data, would a scale of 1 to 7 or 1 to 5 have been sufficient?

3. Field tests of this activity have shown that students are most dissatisfied with facilities and most satisfied with the academics at their institution. Suppose you had switched the order of the categories and presented facilities first. Do you think the changed order would affect the ratings of the academics? What order do you recommend for the questions?

Extensions

Get a copy of the latest *U.S. News and World Report* issue that rates schools. The methodology used to rate the schools is included in the article. You are the consumer, so what variables do you believe the magazine should have used to make the ratings meaningful to you? Write a short recommendation with your suggestions to the magazine.

"What Is a Survey?"

The following was published by the American Statistical Association.

Designing the questionnaire represents one of the most critical stages in the survey development process, and social scientists have given a great deal of thought to issues involved in questionnaire design. The questionnaire links the information needed to the realized measurement.

Unless the concepts are clearly defined and the questions unambiguously phrased, the resulting data are apt to contain serious biases. In a survey to estimate the incidence of robbery victimization, for example, one might want to ask, "Were you robbed during the last six months?" Though apparently straightforward and clear-cut, the question does present an ambiguous stimulus. Many respondents are unaware of the legal distinction between robbery (involving personal confrontation of the victim by the offender) and burglary (involving breaking and entering but no confrontation), and confuse the two in the survey. In the National Crime Survey, conducted by the Bureau of the Census, the questions on robbery victimization do not mention "robbery". Instead, several questions are used which, together, seek to capture the desired responses by using more universally understood phrases that are consistent with the operational definition of robbery.

Designing a suitable questionnaire entails more than well-defined concepts and distinct phraseology. Attention must also be given to its length, for unduly long questionnaires are burdensome to the respondent, are apt to induce respondent fatigue and hence response errors, refusals and incomplete questionnaires, and may contribute to higher nonresponse rates in subsequent surveys involving the same respondents. Several other factors must be taken into account when designing a questionnaire to minimize or prevent biasing the results and to facilitate its use both in the field and in the processing center. They include such diverse considerations as the sequencing of sections or individual questions in the document, the inclusion of check boxes or precoded answer categories versus open-ended questions, the questionnaire's physical size and format, and instructions to the respondent or to the interviewer on whether certain questions are to be skipped depending on response patterns to prior questions.

Source: ASA

Stringing Students Along

Learning about selection bias.

People waiting in a bank queue will need different lengths of time to complete their transactions. For example, depositing a check usually takes less time than obtaining foreign currency. We are interested in finding ways to sample these times so that we might describe their distribution and estimate the average transaction time.

Question

Can I obtain a good estimate of mean transaction time at my bank by averaging the times of all the customers who are at the windows when I arrive?

Objectives

In this activity, you will see a common form of selection bias in a sampling plan that appears to be random.

Prerequisites

You should have some understanding of how to use sample data to estimate population parameters, and you should know how to use a random number table or generator. This should not be your first exposure to sampling.

Activity

The goal of this activity is to estimate the average length of the *N* strings in the bag the instructor has provided. You will sample using two different methods, both involving randomization. The length of each string represents the length of time a customer spends at a bank window.

1. Selection by touch
   a. Without looking into the bag, reach into it and select a string. Record its length (to the nearest inch) and return the string to the bag. Repeat this process until 10 strings have been sampled.
   b. Calculate the average of your sampled string lengths.
   c. Report the average to the instructor.
   d. Repeat the process to get as many additional samples as time allows.
2. Selection by random numbers
   a. Lay the strings out on a table and number them from 1 to *N*.
   b. Select 10 strings by choosing 10 random numbers between 1 and *N* and matching those numbers to the strings.
   c. Measure the length of the sampled strings and calculate the average.
   d. Report the average to the instructor.
   e. Repeat the process to get as many additional samples as time allows.
3. Data analysis
   a. The instructor will give you all the averages recorded in class, for both sampling methods.
   b. Construct a dot plot or stem and leaf diagram of the averages from the sampling by touch.
   c. Construct a dot plot or stem and leaf diagram of the averages from the sampling by random numbers.
   d. Compare your two plots. Do they have the same centers?
   e. Which sampling plan do you think produces the better estimate of the true mean length of the strings? Explain your reasoning.

Wrap-Up

1. a. With regard to the service times at the bank, what method of sampling is represented by the sampling that involves reaching in the bag and selecting 10 strings?
   b. With regard to the service times at the bank, what method of sampling is represented by the sampling by random numbers?
   c. What is the key difference between the observed results for the two different sampling methods?

2. Look at a map of your state, or a neighboring state, that has the lakes marked in a visible way. An environmental scientist wants to estimate the average area of the lakes in the state. She decides to sample lakes by randomly dropping grains of rice on the map and measuring the area of the lakes hit by the rice. Comment on whether you think this is a good sampling plan.

Extensions

1. The sampling by touch method is said to suffer from selection bias in that the longer strings get chosen more often than the shorter ones. What are the implications of this phenomenon in the case of sampling service times at the bank window?

2. Suggest other situations in which selection bias could be important. How might one get unbiased data in these situations?

Assessment Questions

1. You are called upon to help design a sampling plan for the purpose of estimating the average length of stay for patients in a certain hospital. Two options are presented. The first involves randomly sampling patients who happen to be in the hospital the day the sampling is to be done. The second involves randomly sampling names from a list of hospital patients over the past month and checking to determine their records and the length of their stay. Which plan would you choose, and why?

2. Design a sampling plan to estimate the average size of farms in your county.

3. To estimate the average size of classes on a college campus, you could go to the campus, randomly sample some students, and ask them about the sizes of their classes. Or, you could go to the registrar's office and sample from a list of classes being offered. Which method will most likely produce the higher sample mean? Explain why this is true.

4. You randomly choose a sample of people and ask them the number of people who live in their household. The average of these values is *not* a good estimate of the average size of a household. Why not? What *is* this average a good estimate of?

Gummy Bears in Space

Factorial designs and interaction. Controlling variables.

The experimental designs used in this activity are called *factorial designs.* They have two main advantages: They help you study the effects of two or more influences (called *factors*) in a single experiment, and they let you measure the *interaction* between these factors.

Here is an example of interaction from the field of cognitive psychology. How do you answer the question, "Does a cat have claws?" Do you imagine a picture of a cat and check to see whether it has claws, or can you answer without forming a mental image because you associate claws with cats? The psychologist Stuart Kosslyn thought that young children would have to rely on a picture but that older children and adults would have learned the kind of word associations that allow them to answer more quickly, without first forming a picture. To study this theory about language development, Kosslyn used a factorial design.

Factor 1: age groups: (a) first grade, (b) fourth grade, and (c) adult

Factor 2: instructions: (A) imagery instructions—inspect a mental image to answer the question, or (B) no imagery instructions—just answer as quickly as you can

Response: the time it took a subject to answer the question

For his experiment, Kosslyn used all six possible combinations of age group and instructions. The results fit his prediction of an interaction between age and instructions: Adults were slower when they had to call up a mental image than when they did not. First graders were equally slow both ways, suggesting that they used a picture regardless of the instructions.

Question

Your instructor will give you a catapult you will use to launch Gummy Bears. Do the angle of the launcher and its position on the launch ramp interact in their effects on how far the bear goes?

Objectives

In this activity, you will learn how to use a two-way factorial design to study the effects of two factors at once and how to summarize the results in an interaction plot.

Prerequisites

The only true prerequisite is that you should know how to compute and interpret averages. Beyond this, a knowledge of either parallel dot plots or scatter plots will be helpful, but it is not essential. (Because this activity does not depend on measuring variability, you can do it even before you learn about measures of spread.)

Activity

Your instructor will divide your class into teams and provide the equipment you need to launch the Gummy Bears. Imagine that your team is a scientific research group trying to study the effects of two factors on launch distance:

Factor 1: launch angle (the number of books under the end of the launch ramp)

Factor 2: launch position (the launcher is either at the front of the ramp or at the back of the ramp)

1. Plan the experiment.
   a. Each team should assign jobs: (1) loader/launcher, (2) holder (keeps the launch equipment steady), (3) measurer, and (4) data recorder and supervisor.
   b. List the conditions.

      Factor 1: Use three different launch angles: (a) one book, (b) five books, and (c) nine books under the back end of the launch ramp. (The angles, or numbers of books, are called *levels* of the factor.)

      Factor 2: Use two different positions for the launcher on the ramp: (A) front of the ramp, and (B) back of the ramp positions. (This factor has two levels: front and back.)

      For a *two-way factorial* experiment, you *cross* the two factors; that is, you use all possible combinations of the levels of the two factors as conditions for your experiment. In this case, there are six possible combinations of launch angle and launch position.
   c. Randomize the order. Use the six conditions to label six slips of paper and fold the slips so that you can't see the labels. Mix the strips thoroughly and open them one at a time. The first one you draw tells you which condition to use first, and so on. Write the order next to the list of conditions from step 1b.
   d. Keep other conditions fixed. Decide where to position the fulcrum (pencil or dowel) along the bottom craft stick. Mark the place on the craft stick and be sure to check before each launch to make sure you keep the position the same for all your launches.

2. Produce the data. You will record six sets of four launches each, one set for each of the six conditions.

a. Set up your equipment for the first combination of launch angle and launch position on the ramp. Launch a Gummy Bear and measure how far it went. Measure the distance from the *front* of the ramp and measure only in the direction parallel to the ramp, as shown in Figure 1.

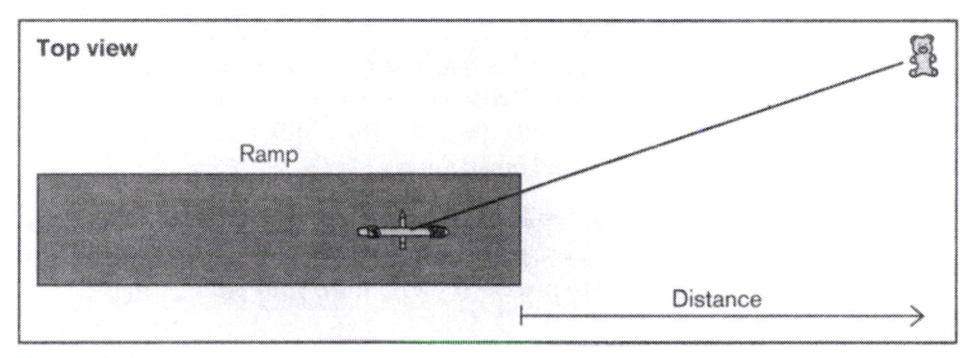

Figure 1

Do three more launches (remember to check the position of the fulcrum!) and record the distance each time. Compute the average of the first four launches.

b. Now reset your launch equipment for the second combination of angle and position on the ramp. Do a second set of four launches and compute their average.

c.–f. Repeat for the third combination of angle and position, and then the fourth, fifth, and sixth.

3. Tabulate the data. Construct a table with two rows and three columns for summarizing your data. Label the rows to correspond to the position of the launcher (front or back) on the ramp. Label the columns to correspond to the number of books (one, five, or nine) under the back of the ramp. Put your average distances in the body of the table.

   As a guide, Table 1 shows actual results from a version of Kosslyn's imagery experiment described at the beginning of this activity. The response times are in milliseconds.

	First Grade	Fourth Grade	Adult
Imagery Instructions	2000	1650	1600
No Imagery Instructions	1850	1400	800

Table 1: Response time

4. Interpret interaction as a "difference of differences." How much distance do you gain or lose by launching from the back of the ramp instead of from the front? Use your data to answer this question by computing the difference (that is, "back average minus front average"). Compute this difference first for the two sets of launches with one book under the ramp. Compute a second difference for the two sets of launches with five books under the ramp and finally compute a third difference for nine books. Show the differences in your table by adding a third row called "Difference." If the two factors (launch angle and position on the ramp) interact, then the three differences will be different. Is interaction present in your launch data?

5. Graph your results. The graphs in Figure 2 show two ways of representing the data from Kosslyn's imagery experiment. Using them as a guide, construct similar graphs to present the data from your table in step 3.

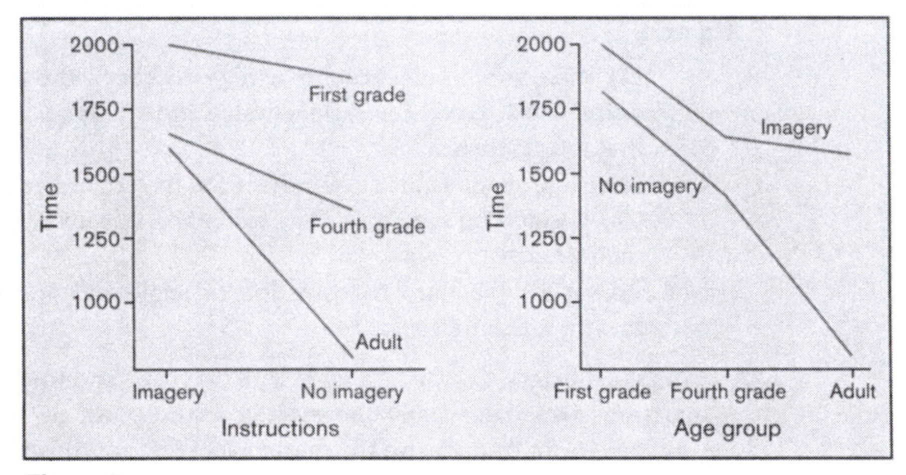

Figure 2

6. Describe the pattern. Write a short paragraph describing the nature of the interaction between the launch position and the launch angle. Your description should tell when you get the longer distances launching from the front position and when you get the longer distances launching from the back position.

Wrap-Up

1. Explain how Kosslyn's results tend to support his theory that word associations like "claws" and "cat" develop later in life than an understanding of the words themselves.

2. Find an example of interaction in a science or social science textbook. List the response and the two interacting factors. Then draw and label a graph that illustrates the interaction you have found.

Extensions

1. Is there another bear-launching interaction? Figure 3 shows three possible positions for the fulcrum of your bear launcher.

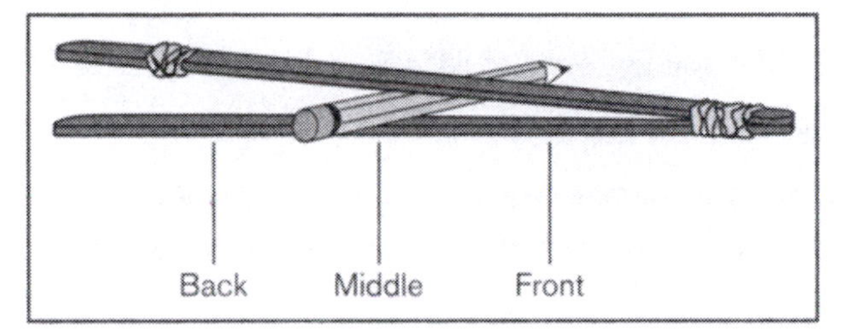

Figure 3

Design and carry out an experiment to test whether there is an interaction between the position of the fulcrum and the launch angle (either one or five books under the back of the launch ramp). As before, launch four times for each combination of conditions and use the average of the four distances as your response value.

2. Let's try a three-factor experiment. (This is an unstructured question, more in the nature of a small project.) Suppose your goal is to find the combination of launch angle, position on the ramp, and position for the fulcrum that will together produce the longest launches. Design an experiment to help you find the best combination of conditions.

Assessment Questions

1. Rectangles

Congratulations! You have just been appointed Director of Research in the Department of Rectangle Science. For your first project, you have designed a two-way factorial experiment:

Factor 1: length—40 or 100

Factor 2: width—10 or 40

Response: area in square inches

a. Make a table summarizing the results you would get from this study. Then draw and label a graph showing the interaction pattern.

b. Clever scientist that you are, you decide to use the same design over again, this time to study the perimeter of the rectangles. Make a table summarizing the results you would get using perimeter as your response. Then draw and label an interaction graph for these results.

c. Write a paragraph comparing your interaction graphs from steps a and b. What is the major difference in the patterns of the two graphs? Why do area and perimeter behave differently?

2. Growth and age
   Imagine a two-way factorial design to study the following scientific hypothesis: "Toddlers get taller; adults don't." Here's a quick summary:

   Factor 1: age groups—2-year-olds and adults

   Factor 2: time—at the start of the study and 3 years later

   Response: height in inches

   Make a two-way table summarizing the results you would expect to get from this study. Then draw and label a graph showing the interaction pattern.

3. Athletic training and heart rate
   Imagine comparing the heart rates of athletes and nonathletes in three conditions: at rest, just after they climb three flights of stairs, and then again after 2 minutes of rest. You will find that on the average that compared with nonathletes, the athletes have a lower resting pulse and that their heart rate increases less during moderate exercise and returns to resting rate more quickly.
   a. Summarize the factorial structure that a study would need to confirm this statement. (List the response and each of the two factors, as in step 1.)
   b. Make a two-way table summarizing the results you would expect to get and show these results in an interaction graph.

4. Warm-blooded, cold-blooded
   a. Complete the following sentence by describing the interaction in words: "If you compare the body temperatures of humans and turtles in the summer when the surrounding temperature is 80°F and in the winter when the surrounding temperature is 40°F, you find that ____________."
   b. Summarize the factorial structure that a study would need to verify the interaction. (List the response and each of the two factors.)
   c. Draw and label an interaction graph showing the kind of results you would expect to get from the study.

5. Thread color and background color
   Design an experiment to study possible interaction between the effects of thread color and background color on the speed of needle threading. Tell how you would measure the response (speed of needle threading) and how you would choose the combinations of conditions to compare.

Funnel Swirling

Experimental design, particularly factorial design. Variability in an experiment.

Consider the following simple game: You will roll a ball bearing down a tube and into a funnel; it will swirl around in the funnel for a while and then roll out onto a table. Your goal is to maximize the amount of time the ball bearing takes to roll down the tube and out of the funnel. You may adjust the height of the funnel and the angle of the tube as you attempt to maximize time.

Question

How good are you at finding the best setting?

Objectives

In this activity, you will learn about factorial experiments and statistical design. You will also study variability and see how it is an inherent part of an experiment.

Prerequisites

You could do this activity on the first day of class, with no statistical preparation. On the other hand, because of the limited number of trials (eight) each group gets with the equipment, groups will benefit from designing their experiments carefully. If you have already studied experimental design, you can put that knowledge to use here.

Activity

Do this activity in teams.

1. Your instructor will divide the class into teams for the "contest" to see which team can keep a ball bearing swirling in the funnel for the longest time. Each team will be given access to a set of materials in order to experiment and collect data.

2. Your team has a budget of $8000. Each trial costs $1000, so you can afford only eight trails—use them wisely.

3. You can adjust two variables during the experiment: (1) the height, *H*, of the funnel (the height of the crossbar) and (2) the angle, *A*, of the tube relative to the crossbar. The height, *h*, of the tube above the crossbar should be 4 inches for each trial. Your team may decide how to conduct your eight trials. You should decide how you will conduct your trials (that is, the settings you will use for each variable for each trial) in advance so that the actual experimentation will go quickly.

4. During a trial, one person should hold the tube and drop the ball bearing, another should hold the funnel steady (if the funnel wobbles it will absorb energy from the ball bearing, which will significantly reduce swirling time), and a third person should operate the stopwatch.

5. After you have completed your eight trials, analyze the data and determine the optimal settings of height and angle. After all teams have conducted their trials and analyzed their data, a "roll-off" will be held to determine the winning team. Each team must submit its "optimal" settings for the variables at the start of the roll-off.

Wrap-Up

1. Did the results of the experiment surprise you? Which factor had the greatest effect on the ball bearing times? What do you now think are the optimal settings of the variables? How do these compare with your original expectations?

2. Write a brief summary of what you learned in this activity about how to design an experiment.

Extensions

Many other variables can be added to the experiment. For example, you could use different ball bearing sizes, with bearing size (say, from 1/4 inch up to 5/8 inch) being a design variable. You could vary the height, *h*, of the top of the tube as a design factor. Another variable is the type of funnel, plastic versus metal. Try repeating the activity while allowing some of these factors to vary.

Assessment Questions

1. Suppose your job is to find the best combination location (in direct sunlight or in shade), watering schedule (once per day or twice per day), and soil pH (low or high) for growing a plant. Describe how you would conduct an experiment to determine the optimal combination.
2. Explain what we mean by an "interaction" in statistics.

Jumping Frogs

Experiments in a factorial situation. Estimating the effects of each factor and of the interaction.

Even the most consistent athletes do not perform exactly the same every time they play. We reason that their performance is affected by several factors, such as the strength of their opponents, their own schedules, or the state of their health. We can use statistics to estimate the effects of such factors on the outcomes. Our sports personality for this activity is going to be a paper frog, and we are interested in the distance it can jump. In particular, we want to find out how the weight and the size of the frog affect "jumping" distance.

We can carry out an experiment to measure the effects of these two factors. The directions for constructing the frogs follow this activity. We have a choice of the type and size of paper to use. We will use two sizes of paper and paper of two different weights. Determining the properties of the ultimate jumping frog is far more complicated and is thus beyond the scope of this activity.

Question

How can we measure one factor's effect on the outcome when we have two factors that can influence the outcome of an experiment?

Objectives

In this activity, you will learn how to analyze an experiment in which the outcomes are affected by two factors, each factor taking on two values. You will learn how to estimate the effect of each factor and the effect of the interaction of the factors on the outcome, without using a formal design and analysis of experiments approach.

Prerequisites

You should be familiar with stem and leaf diagrams and with the use of sample means as estimators.

Activity

1. We want to see how two factors, the size and the weight of the frog, influence "jumping" distance. In this activity, we will use only two sizes and two weights (that is, we will use two levels of each factor):

$$S = \text{size of the frog (2 levels = 2 sizes)}$$
$$W = \text{weight of the paper (2 levels = 2 weights)}$$

   You will use the data on the distances the frogs jump to determine how size and weight affect the frog's jump and use these results to guide you in coming up with the champion frog.
   (Note: Bigger frogs will naturally weigh more than smaller ones. But in this experiment, the term *weight* refers only to the "weight" of the paper. You can have small frogs made of heavy paper that may, in fact, weigh less than large frogs made of light paper.)

2. You will need square sheets of paper to construct your frog. Your instructor will have 7.5-inch and 6-inch squares in two weights of paper. Each of you will be randomly assigned one of the squares. Make a frog with your sheet of paper. Practice with your frog five times and report to the instructor the length of the last jump. (Measure the jump in centimeters.)
   Enter the data for the class in Table 1.

		Size	
		Large	Small
Weight of Paper	Heavy		
	Light		

Table 1: Jumping distance

3. We will first look at the results of our experiments graphically.
   a. Construct a back-to-back stem and leaf diagram with all measurements for the large paper frog jumps on one side and those for the small paper frog jumps on the other side. Note that you are using both heavy and light paper measurements on each side. You may wish to show them with different symbols.
   b. Repeat the back-to-back stem and leaf diagram with heavy paper frog jumps on one side and light paper frog jumps on the other side. Again, you could indicate the large paper and small paper frog jumps with different symbols.
   c. What do the stem and leaf diagrams tell us about the relative distances jumped by large and small frogs? Are there differences in the patterns of jumps of the heavy and light frogs in the first back-to-back stem and leaf diagram?

d. Now turn your attention to the second back-to-back stem and leaf diagram. Compare the distances jumped by the heavy and light frogs in the second stem and leaf diagram. Does the size of the frog seem to affect the jump?

e. Combine your analyses to reach a conclusion on which factor has more of an effect on the jumps: size or weight.

4. We will now construct arithmetic summaries for the data. We will try to get an estimate of the effect of size and weight on a frog's jump. Let's look at the following table of means, Table 2.

		Size	
		Large (+)	Small (−)
Weight of Paper	Heavy (+)	Mean of heavy/large frogs = W_+S_+	Mean of heavy/small frogs = W_+S_-
	Light (−)	Mean of light/large frogs = W_-S_+	Mean of light/small frogs = W_-S_-

Table 2: Mean jumping distance

You now need to calculate the mean of all the heavy frogs and that of all the light frogs by getting the two row averages. So, for example, the mean jump for the heavy frogs is $1/2[(W_+S_+) + (W_+S_-)]$. Based on these averages, which weight gives better jumping frogs? To estimate the difference in the length of the jumps due to the different weights, calculate

$$A = 1/2(\text{mean jump of the heavy frogs} - \text{mean jump of the light frogs})$$

We will use this as an estimate of the effect of the weight of the paper. Now calculate the mean of all the large frogs and that of all the small frogs by calculating the column averages. Which size frog jumps better? The effect of the paper size is given by

$$B = 1/2(\text{mean jump of the large frogs} - \text{mean jump of the small frogs})$$

In the previous section, you reached a conclusion on the relative effects of weight and size. Do the values of A and B confirm it? Using the values of A and B, complete the following statements:

Changing the weight of the paper (frogs) from light to heavy (increases/decreases) the average distance jumped by ____________.

Changing the size of the paper (frogs) from small to large (increases/decreases) the average distance jumped by ____________.

5. Do the two factors interact with each other? In other words, is the difference in the mean jumps of the large frogs and that of small frogs different for the heavy paper and the light paper? We can estimate the interaction term as follows. First, for the large paper frogs, we calculate the average difference due to weight as

$$D1 = 1/2(\text{mean of the large/heavy} - \text{mean of the large/light})$$

and for the small frogs we get

$$D2 = 1/2(\text{mean of the small/heavy} - \text{mean of the small/light})$$

The quantity $1/2(D1 - D2)$ gives us a measure of the interaction term or the effect of changing the levels of both factors. Looking at the table of means and the estimate of the interaction between the two factors, are the differences in the average jumps between large and small frogs the same for both weights of paper?

Wrap-Up

1. We were able to estimate the effect of the two factors using means. Do the following graphical analysis: Find the overall mean. Complete the following graph using the row and column means from Table 2. Above the label "Size" you will plot the two column means, and plot the two row means above the label "Weight." Use the appropriate vertical scale.

   Look at the distances of the two means for size from the overall mean. Are the two means equidistant from the overall mean? Compare the magnitude of this distance to the value of B that you computed earlier. Do a similar analysis on the distances of the two means for weight from the overall mean, comparing them with the value of A. Does it seem reasonable to use A and B as measures of the effects of the two factors? Comment.

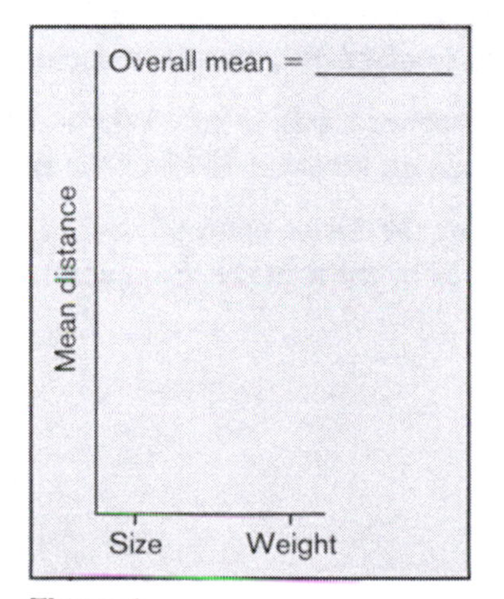

Figure 1

2. Complete the following graph using the means W_+S_+, W_-S_+, W_+S_-, and W_-S_- from Table 2. You will plot W_+S_+ and W_-S_+ above "Large," and W_+S_- and W_-S_- above "Small."

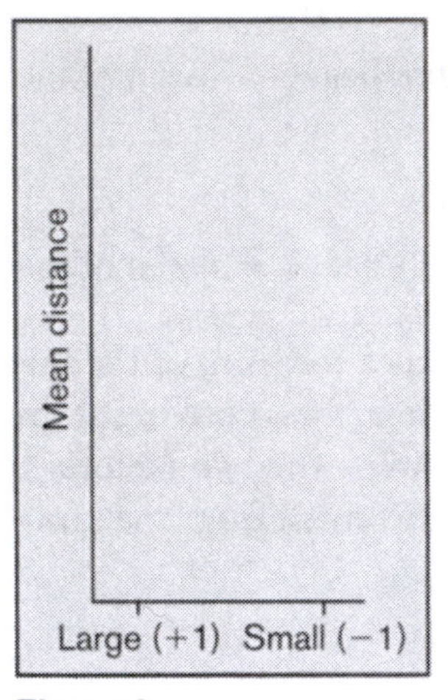

Figure 2

Now join the points W_+S_+ and W_+S_-. Also join the points W_-S_+ and W_-S_-. You now have two lines. Compare the slopes of the two lines. Are the lines parallel? If we replace "Large" and "Small" on the horizontal axis by $+1$ and -1, respectively, we can get the slopes of the two lines. Calculate the slopes and compare them with *D*1 and *D*2 that you calculated earlier. Explain how the measure for interaction of the two factors is related to the slopes of the two lines.

Extensions

This activity is an analysis of data from a designed experiment. The response was affected by two factors, the size and the weight of paper, each factor taking on two values. We estimated the effects of the size and the weight of paper used on the jumps of the frogs as well as the interaction of these two factors.

1. Your instructor randomly assigned a frog to you. Why was this done? Give examples of how nonrandom assignments could affect the analysis.

2. Write a brief report describing the design and analysis of an experiment to find out how test scores are affected by the time the test is administered and the level of difficulty of the test.

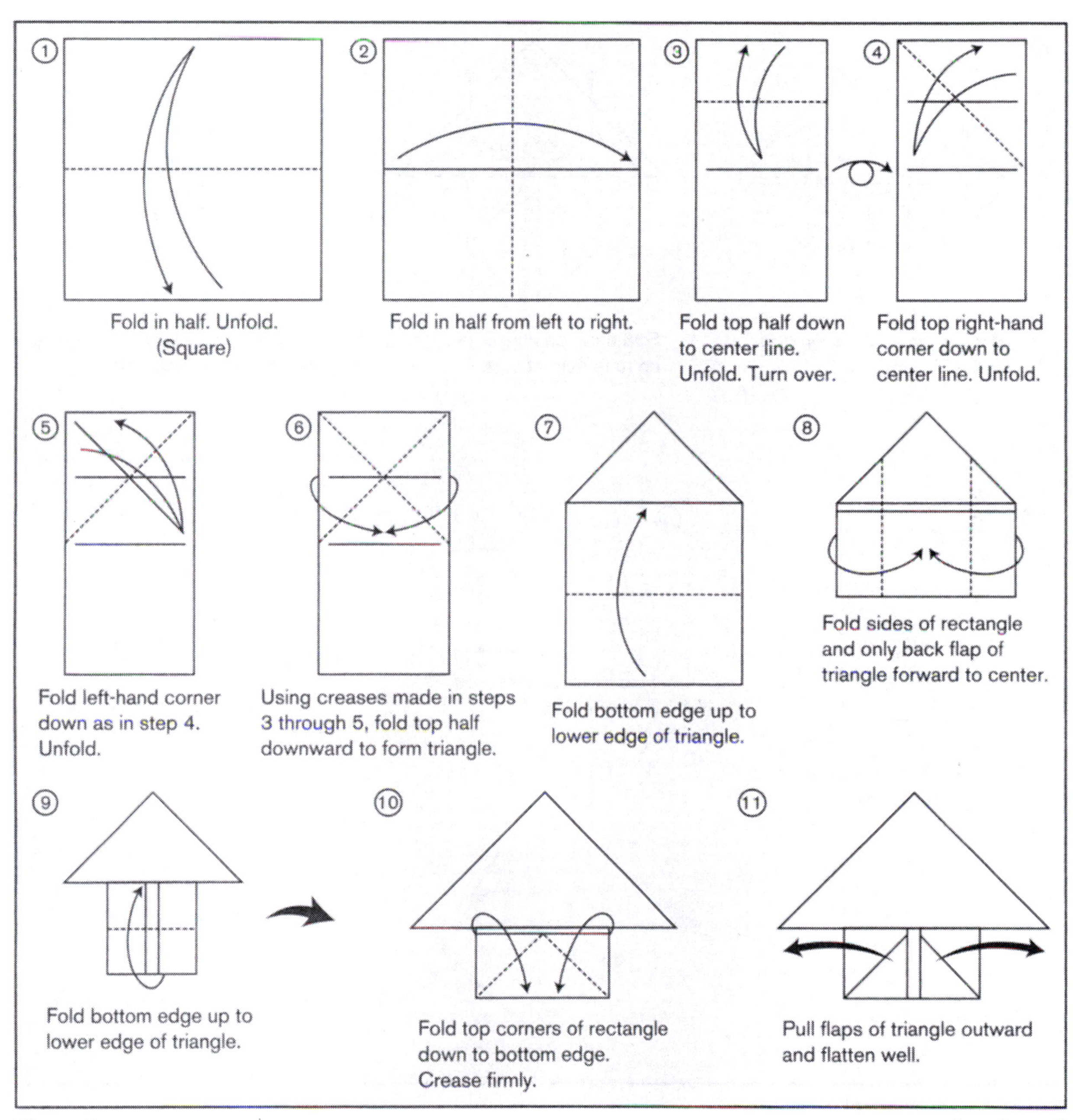

Figure 3 ***(continued)***

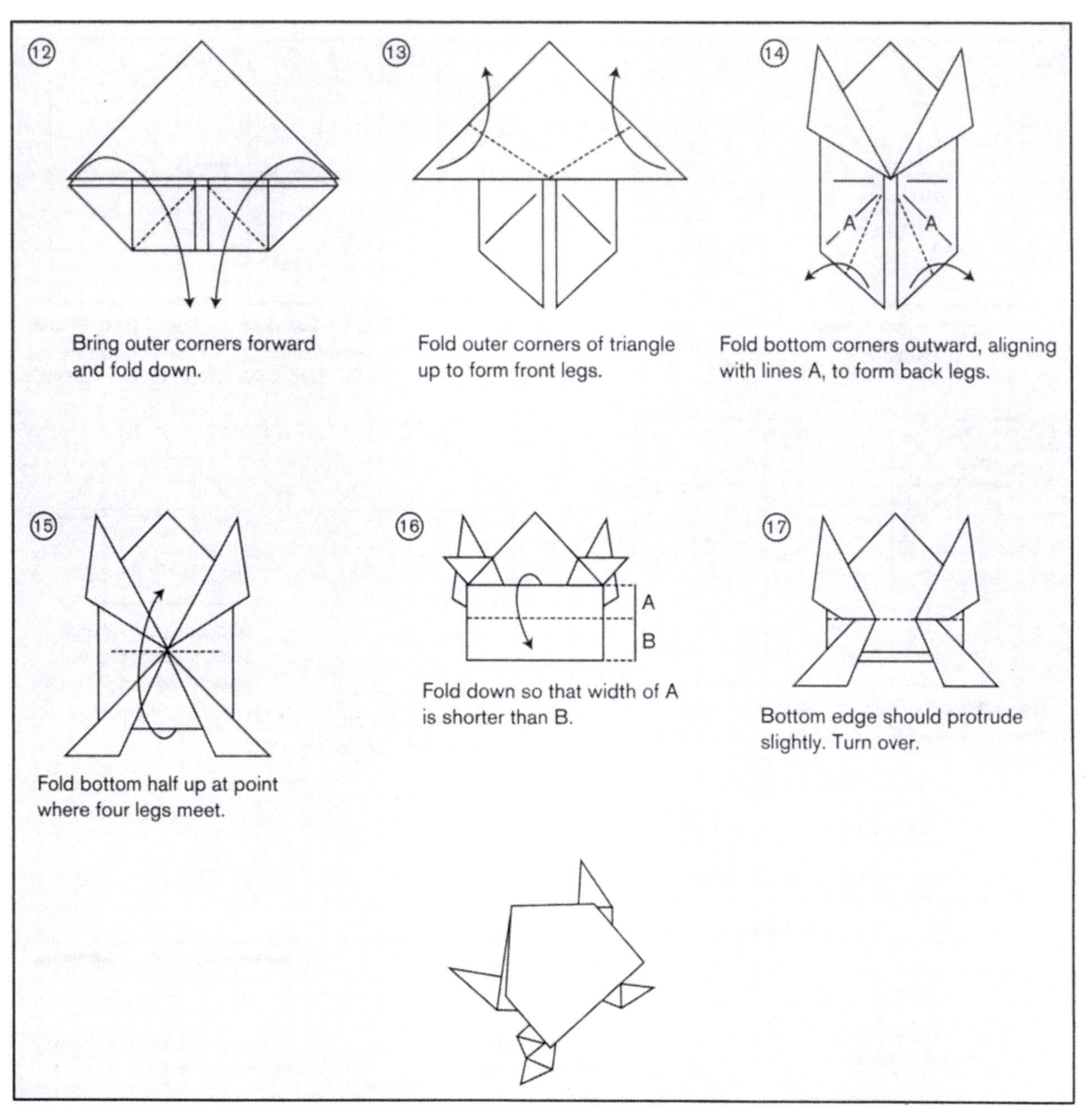

Figure 3 ***(continued)***

Assessment Questions

1. Why do you suppose the "effect" sizes *A* and *B* have a (1/2) in them?
2. Describe a real-life situation analogous to the frogs—two variables, two levels each—that would have a large interaction term.

How to Ask Questions: Designing a Survey

How the way a question is worded can affect the outcome of a survey.

Political polls do a remarkably good job of predicting the winner of national elections. But sometimes a poll goes badly wrong. An example was the 1992 election in Colorado, where a measure was on the ballot to prohibit the state legislature and cities from passing antidiscrimination laws concerning homosexuals. Polls showed that the Colorado measure would be defeated. The measure passed. (See the *New York Times*, November 8, 1992.)

Question

What may have accounted for the difference between the result in the polls and the result in the election?

Objectives

In this activity, you will examine how the design of a survey can affect the responses.

Prerequisites

You should be familiar with the hypothesis test for the equality of two population proportions.

Activity

With your group, select one of the following questions. Then design and conduct a simple experiment to answer it. That is, you will design a survey and administer it to some people. The point will be to study the effect of how you ask the question, not to determine its answer. Use the students at your institution as the population of interest.

Before doing the survey, your group will have to decide on a sample size that will be large enough to establish the statistical significance of any difference you feel is meaningful.

1. Does the order in which two candidates appear on a ballot make a difference in the percentage of votes they receive?
2. Is it possible to word a question in two different ways that are logically equivalent but that have a different percentage of students agree with them?
3. Does the order in which two statements appear in a survey make a difference in the percentage of students who agree with them?
4. Can the percentage who agree with a statement be changed by having respondents read some introductory material?
5. If a statement is rewritten to be logically equivalent but to have a more complicated sentence structure and bigger words, will it affect the percentage of students who agree with it?
6. Does the appearance of the interviewer make a difference in how students will respond to a question? For example, do students tend to respond the same way about a controversial issue when the interviewer is female as when the interviewer is male?
7. If the interviewer does not know how a student responds (as on a secret ballot), does it make a difference in the percentage of students who agree with a controversial statement?
8. If a student knows absolutely nothing about an issue, will he or she give an opinion anyway? Will students admit it if they don't know the answer to a question?
9. Do students report events (such as how many days last week it rained or the description of a person who just walked by) as accurately as they think they do?
10. If you let students volunteer to be in your poll, do you get a different result than if you approach the students?

Wrap-Up

1. Tabulate the data from your survey. As part of a report of your findings, do any relevant statistical tests (such as a Z test for a difference in proportions).
2. Your group should prepare a report about what you have learned and present it to the rest of the class. Remember that the point of your report is primarily to describe the effect of the *design* you are studying, not the students' opinions about the issue.
3. Write a set of guidelines that the school paper might use to conduct surveys of student opinion.

III

Anticipating Patterns

What Is Random Behavior?

Gambler's fallacy. It's hard to predict short-term random behavior.

A basketball player who makes 70% of her free throws has missed four in a row. As she is about to take her fifth free throw, a spectator says that she is "due" to make this one. You have lost card games to your friends every night this week; you think to yourself, "Well, next week I'm due to win a few." A couple has three children, all boys. "The next one is due to be a girl," a friend remarks. Each of these events, the free throw, the card game, the birth of a child, involves a chance outcome that cannot be controlled. In the long run, the basketball player makes 70% of her free throws, you may win 30% of the card games, and approximately 50% of the babies born are girls. But do these long-run facts help us predict outcomes of random events over the short run?

Question

Can knowledge of past events help us predict the next outcome in a random sequence?

Objectives

In this activity, you will try to predict the next outcome in a random sequence of events based on your knowledge of the past outcomes in the sequence. Because the sequence is random, your predictions will not always be correct, but some prediction methods may be better than others.

Prerequisites

You should have some knowledge of the relative frequency interpretation of probability.

Activity

1. The large box of beads in the classroom contains some that are yellow (or otherwise identifiable; it will depend what your instructor provides). Beads are to be selected from the box one at a time and laid on the desk in the order of selection. Your goal is to come up with a rule for predicting the occurrence of yellow beads. To begin, you'll use a sample of five beads from the box to serve as data for making the first prediction.

2. Construct a rule for predicting whether the next bead drawn will be yellow. You may use the information on the first five selections, but your rule must work for any possible sample selection of beads.

3. Record the prediction your rule generates for whether the next bead will be yellow. Now, randomly select the next bead. Record whether your prediction was correct.

4. Continue making predictions for the next selection and then checking them against the actual selection for a total of at least 25 selections (not including the first 5). Keep using your same rule. Calculate the percentage of correct predictions your rule generated.

5. Compare your rule with others produced in the class. Arrive at a class consensus on the "best" rule for predicting yellow beads.

6. Do the data from past selections help in making a prediction for the next selection? Does the particular pattern of the sequential selections help in making a prediction for the next selection? What is the best you can hope to do, in terms of correct predictions, with any decision rule? Discuss what was learned about predicting future outcomes in a random sequence.

Wrap-Up

1. Discuss the three situations posed at the beginning of this activity. In light of your experience with random sequences, are the statements made about the free throws, card games, and sex of newborns appropriate?

2. Repeat steps 1 through 6 of this activity using the *same* random device but generating a new and different sequence of outcomes. Compare all of the prediction rules once more. Did the percentages of correct predictions remain the same as before for all rules? Did the rule originally chosen as best remain the best decision rule?

3. Repeat steps 1 through 6 of this activity with another random device generating an outcome with probability closer to 0.5. Will your ability to predict get better? Why or why not?

Extensions

1. Picture a square table for four people, with one chair on each side of the square. Two people come to the table to eat together. Do they sit beside each other or across from each other?
   - **a.** Construct a rule for predicting whether two people sitting down at a square table will sit across from each other.
   - **b.** Go to a cafeteria that has a number of square tables that seat four to check out your rule. What fraction of the time were you correct?
   - **c.** Under the assumption that people choose seats at random, what fraction of the time should two people sit across from each other? How does this fraction compare with your answer to 1b?
   - **d.** Based on your observations, is the pattern of seating consistent with the assumption that people choose seats at random? Based on your life experience, do you think people choose seats at random?

2. You are playing a card game in which some of the cards are dealt face-up. Using a standard 52-card bridge deck, the chance of a face card on a single draw is 12 in 52, or about 0.23. So if we predict that the first card will not be a face card, we will be right 77% of the time. After the game has progressed awhile, can you predict whether the next card will be a face card with a rule that is accurate more than 77% of the time? Explain.

Assessment Questions

1. Suppose we know that 70% of the students on a large campus are male. Random drawings are to be made to give away five tickets to a concert. Can you predict whether the first ticket given will go to a male? Can you do any better in predicting whether the fifth ticket will go to a male, given that the first four have already gone to males?

2. Discuss how the activity on predicting outcomes of random sequences relates to the process of conducting an opinion poll. Does the long-run relative frequency of responses help us predict how the next person interviewed will respond? Does the short-run relative frequency of responses help us predict how the next person will respond?

3. Find articles from the media in which the "short-term" prediction scheme is used as an argument. In some of these, such as games of chance, the past does not help predict the future. In others, such as rain after a dry spell, the past might provide some information on the future. Form clear arguments about which cases look like the random sequence activity and which do not.

4. An entrepreneur is selling advice on how to choose numbers to improve your chances of winning a lottery. If he approaches you as a potential customer, what would you say?

The Law of Averages

What is the law of averages? How probability helps us predict in the long term.

Have you ever heard someone appeal to the "law of averages"?

- As many times as she's been guilty before, she has to be innocent this time—it's the law of averages. You should know that. You want to go to law school. (Tom on *The Tom Arnold Show,* March 9, 1994.)
- President Reagan explained why there is corruption in the Pentagon's purchasing process: "The law of averages says that not all 6 million [people involved] are going to turn out to be heroes." (*Pittsburgh Press,* June 18, 1988.)
- Disobey the law of averages. Let others take the traditional course. We prefer creativity over conformity. Invention over imitation. Inspired ideas over tired ideas. In short, Audi offers an alternative route. (From a print advertisement for Audi.)

Question

What do people mean when they refer to the "law of averages"?

Objectives

In this activity, you will examine how people use the term "law of averages" to make correct and incorrect inferences about probability. Then you will explore the consequences of the law of large numbers, a correct statement about probability.

Prerequisites

None.

Activity

1. There are several popular versions of the law of averages. Cut apart the quotations listed in "Statements about the Law of Averages" (pages 122–124) and sort them into piles according to what the author or speaker meant when using the phrase the "law of averages." For example, one pile can include statements trying to convey that, given enough opportunities, even unlikely events must eventually happen. The Reagan statement would go into this pile. A second pile can include statements that contend that if an event has not happened on several previous opportunities, it is much more probable that it will happen on the next opportunity. Tom Arnold's statement would go into this pile. A third pile can include statements that use the following definition:

   Law of averages: the proposition that the occurrence of one extreme will be matched by that of the other extreme so as to maintain the normal average. (From *Oxford American Dictionary,* 1980.)

   Miscellaneous uses, such as the Audi ad, might go into a final pile.

2. Which of your categories of the "law of averages" are correct interpretations of probability?

3. Following is a sequence of flips of a fair coin:
   T T H H T H T T H T H H T T T H T H H H H T T T H H T H H H H H T H
   T H T T H H H T H T H H T H H T
   a. How many heads would you expect to have after 10 flips of a fair coin? ("Expect" is a technical term that means about the same as "average." This question means, "On the average, how many heads do people get after flipping a coin 10 times?")
   b. In the preceding sequence, what is the actual number of heads after 10 flips? What is the percentage of heads after 10 flips?
   c. How many heads would you expect to have after 50 flips of a fair coin?
   d. In the preceding sequence, what is the actual number of heads after 50 flips? What is the percentage of heads after 50 flips?

   What many people find counterintuitive about the answers to step 3 is that the percentage of heads gets closer to 50% while the actual number of heads gets further away from half the total number of flips. In fact, the law of large numbers tells us that this is what we can expect.

4. You will win a prize if you toss a coin a given number of times and get between 40% and 60% heads. Would you rather toss the coin 10 times or 100 times, or is there any difference?

5. You will win a prize if you toss a coin a given number of times and get exactly 50% heads. Would you rather toss the coin 10 times or 100 times, or is there any difference?

6. Suppose you plan to flip a coin indefinitely. The first two flips are heads.
   a. Is a head or a tail more likely on the next flip?
   b. How many heads do you expect to have at the end of 10 flips? What is the expected percentage of heads at the end of 10 flips?
   c. How many heads do you expect to have at the end of 1000 flips? What is the expected percentage of heads at the end of the 1000 flips?
   d. As you keep flipping the coin, what happens to the expected number of heads? What happens to the expected percentage of heads?

Wrap-Up

1. Does the law of large numbers imply that if you toss a coin long enough, the number of heads and the number of tails should even out? Explain.

2. As Los Angeles Laker Carl Perkins comes up for a seventh free throw against the Golden State Warriors, announcer Chick Hearn notes that Perkins had made the last six out of six free throws and concludes that "the law of averages starts working for Golden State" (December 15, 1990). What does Chick Hearn mean? Is this a correct interpretation of probability?

Extensions

1. Design a survey to determine what a typical person thinks the phrase "law of averages" means.

2. This question is adapted from D. Kahneman et al. (1982), *Judgment under Uncertainty: Heuristics and Biases*, Cambridge: Cambridge University Press.
   a. The average math achievement test score of the population of eighth graders in a large city is known to be 100. You have selected a child at random. Her score turns out to be 110. You select a second child at random. What do you expect his or her score to be?
   b. The average reading achievement test score of the population of eighth graders in a large city is known to be 100. You have selected 50 children randomly. The first child tested has a score of 150. What do you expect the mean score to be for the whole sample?

3. Design an experiment to test this version of the law of averages: In gambling games, if an event hasn't happened at the last few opportunities, it is more likely to happen at the next opportunity.

Assessment Questions

1. A roulette wheel has 18 black positions and 18 red positions. A gambler observes six consecutive reds and then bets heavily on black because "black is due." Is his reasoning correct? Explain.

2. After hearing your explanation to question 1, the gambler moves on to a poker game. He is dealt four red cards. He remembers what you said and assumes that the next card dealt to him is equally likely to be red or black. Is the gambler right or wrong? Why?

Statements about the Law of Averages

Morse, too, was beginning to wonder whether he himself was following the drift of his own logic, but he'd always had the greatest faith in the policy of mouthing the most improbable notions, in the sure certainty that by the law of averages some of them stood a more reasonable chance of being nearer to the truth than others. So he burbled on. [C. Dexter (1983), *The Riddle of the Third Mile,* New York: Bantam, p. 110.]

By the law of averages, it is implausible that these two people would ever come in contact with each other. (FM radio program, March 5, 1992.)

Dear Abby: My husband and I just had our eighth child. Another girl, and I am really one disappointed woman. I suppose I should thank God she was healthy, but, Abby, this one was supposed to have been a boy. Even the doctor told me that the law of averages was in our favor 100 to one. [Abigail Van Buren, 1974. Quoted in Karl Smith (1991), *The Nature of Mathematics,* 6th ed., Pacific Grove, CA: Brooks/Cole, p. 589.]

According to the law of averages, every so often a perfect Mac rolls off Apple's highly automated assembly line. (*MacWorld Magazine,* April 1994, p. 153.)

There are two major problems with the law of averages. First, it does not exist. What sportscasters think they're talking about is called the Law of Large Numbers. Second, they use this law when it does not apply. It has absolutely no application in sports.

The Law of Large Numbers states that if you know the final average of a large sequence of numbers, and the average after a large number of observations is either above (or below) the final average, then each subsequent observation is more likely to be below (or above) the final average. If you think about it, this makes sense. Say you have a series of a million numbers, and you know the average of these numbers is 10. If, after 800,000 observations, the average is 9.5, then each subsequent observation is more likely to be above 10. (*Streaks, Slumps and the Law of Averages: Recommended Reading for Sportscasters.* Iain Fyffe, 2002 at http://www.hockeyzoneplus.com/puckerings/puck004.htm.)

It seems that the media are constantly feeding us with statistics. One in four people are going to get this or one in five people have got that. They tell us that one in three marriages are going to end in divorce. They say that the average person consumes "x" litres of wine each year or the average person drinks so much tea a year. The media are using what is called the law of averages. The average is the arithmetical mean, established by adding a set of values and dividing the total by the number of items. It represents the middle or median point in a range of given values. In these days most database programs provide the means to establish the average of almost everything. Many people believe what the media says, and they believe the law of averages. (*How to Outdo the Law of Averages* by Peter Wade.)

In Vinnie's own country [the United States], according to statistics (borne out by her own observation) one out of three men over thirty is overweight. Here [England] most remain trim; but those few who do become fat, as if by some law of averages, often become excessively so. [Alison Lurie (1984), *Foreign Affairs,* New York: Avon, p. 245.]

"The law of averages is what baseball is all about," says Kiner. "It is the leveling influence of the long season. A .250-hitter may hit .200 or .300 for a given period. But he will eventually level off at .250. The same is true of pitchers. Illnesses, sore arms, good and bad clubs are all part of it. But the law is inflexible. A player will average out to his true ability."

What this means in Seaver's case is that he is now "paying" in the percentages for his 1969 season in which he had a 25–7 record. . . . ["Baseball law of averages taking toll on Seaver," *New Haven Register,* June 2, 1974. Quoted in Gary Smith (1985), *Statistical Reasoning,* Boston: Allyn and Bacon, p. 175.]

This pitcher has won his last ten games, so if you think about the law of averages. . . . But every time you say something like that you look in the record book and find some guy with eleven straight wins. (Paraphrase of Dodger announcer Vin Scully, October 1, 1991.)

"What I think is our best determination is it will be a colder than normal winter," said Pamela Naber Knox, a Wisconsin state climatologist. "I'm basing that on a couple of different things. First, in looking at the past few winters, there has been a lack of really cold weather. Even though we are not supposed to use the law of averages, we are due," said Naber Knox, an instructor in meteorology at the University of Wisconsin—Madison. (Associated Press, Fall 1992.)

There *is* a fall [season] here. It's not as dramatic as in other places, but it's the only fall we have. And the law of averages dictates that you'll need some new clothes to get through it. (*Los Angeles Times Magazine,* September 26, 1993, p. 30.)

Grover Snodd was a little older than Tim, and a boy genius. Within limits, anyway. A boy genius with flaws. His inventions, for example, didn't always work. And last year he'd had this racket, doing everybody's homework for them at a dime an assignment. But he'd given himself away too often. They knew somehow (they had a "curve," according to Grover, that told them how well everybody was supposed to do) that it was him behind all the 90s and 100s kids started getting. "You can't fight the law of averages," Grover said, "You can't fight the curve." [Thomas Pynchon (1984), *Slow Learner: Early Stories,* Boston: Little, Brown, and Company, p. 142.]

Boy's official attitude to the run-through of the last act was that since the first three had gone so abysmally, by the law of averages, the fourth act must go better. [Antonia Fraser (1982), *Cool Repentance,* London: Methuen, p. 144.]

He had often wondered how many people died in hotels. The law of averages said some would, right? [Anne Tyler (1985), *The Accidental Tourist,* New York: *Berkley,* p. 331.]

The period of the big parties had been the critical time. Making scores of undistinguished bohemian people interesting to each other, imbuing them with poise, confidence, and urbanity for a night, had been stiff labour; and even on unauthentic champagne-cup and mixed biscuits it had come expensive. But again it worked. A mere law of averages mingled in those promiscuous gatherings certain of the emerging Great. [Michael Innes (1937), *Hamlet, Revenge!,* London: Penguin Books.]

Streaky Behavior

Runs in Bernoulli trials. Randomness is streakier than we think.

When they have made several baskets in succession, basketball players are often described as being "hot." When they have been unsuccessful for a while, they're said to be "cold" or "in a slump." Fans and basketball players alike tend to believe that players shoot in streaks. That is, players have long periods when they are shooting better than we would expect followed by long periods when they aren't doing as well as we would expect. In this activity, you will begin to evaluate whether it is true that basketball players exhibit streaky behavior in shooting. The first step is to learn to recognize a streak of unusually successful or unsuccessful shooting.

Question

How can we recognize a streak of unusually successful basketball shooting?

Objectives

In this activity, you will study the distribution of the length of the runs of successes in a sequence of Bernoulli trials. (A Bernoulli trial, also called a *binomial trial,* is one of a sequence of independent trials, where each trial has two possible outcomes—usually called *success* and *failure.* If the probability of success is p, the probability of failure is $1 - p$.) The probability, p, must remain constant across all trials.

Prerequisites

None.

Activity

1. Without using a coin or a random number table, write down a sequence of heads (H) and tails (T) that you think resembles the results that one would get from actually flipping a coin 200 times.

2. In the following two sequences, H stands for a head and T for a tail. One of the sequences is the result from actually tossing a fair coin. The other was made up by a person. Which one do you think is from the coin?

   THTTTTHHTHTHTTTTTHHTHTHHHHHHHHHTHTHTHHHT
   HHHTTHHTHTHHHHHTHHTHHTTHTHTHTTTTTHHTHTH
   THTTTHTTTTHHTTTTHTTTTHTTTTHHHTHTHHTTTHTT
   THTTHHTHHHTHHHHHTHHHTTHHHTHHTTTTTTTHTTTH
   TTTHTTHTTTHTTHHHHTHHHTTTHTTTTTTHHTHTTTHH

   THTHTTTHTTTTHHTHTTTHTTHHHTHHTHTHTHTTTTHH
   TTHHTTHHHTHHHTTHHHTTTHHHTHHHHTTTHTHTHHHH
   THTTHHHTHHTHTTTHHTHHHTHHHHTTHTHHTHHHTTTH
   THHHTHHTTTHHHTTTTHHHTHTHHHHTHTTHHTTTTHTH
   THTTHTHHTTHTTTHTTTTHHHHTHTHHHTTHHTHHTHHT

3. How many from your class chose the first sequence? The second? What were the reasons for the choices?

4. Construct a real sequence of 200 coin tosses by tossing a coin or using a random number table. What is the longest run of heads in your sequence?

5. Make a frequency table of the lengths of the longest runs of heads from your class's sequences. From the frequency distribution, make a plot of the lengths. Compute an estimate of the mean length of the longest run of heads in a sequence of 200 coin tosses. Mark this mean on the histogram.

6. Which sequence in step 2 do you now think is the real one? Explain.

7. Does your impostor sequence from step 1 resemble a real sequence of coin tosses? In what way is it different?

Wrap-Up

1. If, on average, a basketball player makes 50% of field goals attempted and shoots 200 times in a series of games, what would you expect the longest streak of baskets to be if each shot is a Bernoulli trial? How could you tell if the player was "hot?"

2. What could be some of the reasons people tend to believe in streak shooting?

3. Would a frequency table of the lengths of the longest run of *tails* be the same as or different from that of the longest run of heads? Explain your reasoning.

Extensions

1. Pick a basketball player whom you can follow for a few games and who has a well-established field-goal percentage. Using this percentage and random numbers, make a frequency table of the longest run of heads in a sequence of 200 attempts. After following the player for 200 field-goal attempts, do you see any evidence of an unusually long streak?

2. Construct the probability distribution for the length of the longest run of heads when a coin is tossed one, two, three, four, and five times. For example, when a coin is tossed two times, the possible outcomes are HH, HT, TH, and TT. The longest runs of heads in these outcomes are 2, 1, 1, and 0, respectively. The probability distribution for the longest run of heads is shown in Table 1.

Longest Run of Heads	Probability
0	1/4
1	2/4
2	1/4

Table 1

After examining your probability distributions, what conjectures can you make?

Assessment Questions

1. Estimate the probability that there will be eight or more heads in a row when a coin is flipped 200 times.

2. In this activity, you constructed a histogram of the longest run of heads in a sequence of 200 coin flips. The probability of a head was 0.5 on each flip. Suppose instead you were constructing a histogram of the longest run of fives in a sequence of 200 tosses of a die for which the probability of a five is 1/6 on each toss. How would this new histogram be different from the one you constructed? How would it be the same?

3. A book review in the *Los Angeles Times* of December 29, 1993, about discrimination against upper-class African Americans contains the following example:

A Harvard law student says her professor tended to ignore the raised hands of the black students in class—and then, suddenly, he would call on several black students in a row: "As if," she explains, "the professor had suddenly realized that he was neglecting an important segment of the student body and had resolved to make amends."

Discuss this quotation in the light of what you have learned from this activity.

Counting Successes

How to create simulations to study problems about the number of successes in repetitions of an event with a known probability.

A basketball player goes to the free-throw line 10 times in a game and makes all 10 shots. A student guesses all the answers on a 20-question true/false test and gets 18 of them wrong. Of 12 persons selected for a jury, 10 are female. Unusual? Some would say that the chances of these events happening are very small. But, as we have seen, calculating the probabilities to evaluate the chances of these events requires some careful thought about a model. These three scenarios all have certain common traits. They all involve the repetition of the same event. They all have as a goal counting the number of "successes" in a fixed number of repetitions. This activity discusses how to construct a simulation model for events of this type so that we can approximate their probabilities and decide for ourselves whether the events are unusual.

Question

How can we simulate the probability distribution of "the number of successes out of n repetitions"?

Objectives

In this activity, you will learn to recognize probability problems with outcomes of the form "number of successes in n repetitions of an event," to simulate the distribution of such outcomes, and to use simulated distributions for making decisions.

Prerequisites

Some elementary knowledge of fractions, percents, and proportions.

Overview

A key to understanding simulations is to follow a standard procedure.

1. First, identify the basic random component and find a device that will generate outcomes with this probability.
2. Second, understand how many of these basic components are necessary for one trial of the simulation (1 in the case of purchasing one soft drink, 10 in the case of taking a 10-question multiple-choice test).
3. Third, identify what is being counted as a success and accumulate the total number of successes across the trial. This results in one random outcome.
4. Finally, repeat the whole procedure many times to generate a simulated distribution of outcomes.

Activity

1. Read the following article from the *Milwaukee Journal* (May 1992) titled "Non-cents: Laws of probability could end need for change."
   a. Does this seem like a reasonable proposal to eliminate carrying change in your pocket?
   b. Do you think the proposal is fair? Explain your reasoning.

Non-cents: Laws of Probability Could End Need for Change

Chicago, Ill.—AP—Michael Rossides has a simple goal: to get rid of that change weighing down pockets and cluttering up purses.

And, he says, his scheme could help the economy.

"The change thing is the cutest aspect of it, but it's not the whole enchilada by any means," Rossides said.

His system, tested Thursday and Friday at Northwestern University in the north Chicago suburb of Evanston, uses the law of probability to round purchase amounts to the nearest dollar.

"I think it's rather ingenious," said John Deighton, an associate professor of marketing at the University of Chicago.

"It certainly simplifies the life of a businessperson and as long as there's no perceived cost to the consumer it's going to be adopted with relish," Deighton said.

Rossides' basic concept works like this:

A customer plunks down a jug of milk at the cash register and agrees to gamble on having the $1.89 price rounded down to $1 or up to $2.

Rossides' system weighs the odds so that over many transactions, the customer would end up paying an average $1.89 for the jug of milk but would not be inconvenienced by change.

That's where a random number generator comes in. With 89 cents the amount to be rounded, the amount is rounded up if the computerized generator produced a number 1 to 89; from 90 to 100 the amount is rounded down.

Rossides, 29, says his system would cut out small transactions, re-

(continued)

ducing the cost of individual goods and using resources more efficiently.

The real question is whether people will accept it.

Rossides was delighted when more than 60% of the customers at a Northwestern business school coffee shop tried it Thursday.

Leo Hermacinski, a graduate student at Northwestern's Kellogg School of Management, gambled and won. He paid $1 for a cup of coffee and a muffin that normally would have cost $1.30.

Rossides is seeking financial backing and wants to test his patented system in convenience stores.

But a coffee shop manager said the system might not fare as well there.

"Virtually all of the clientele at Kellogg are educated in statistics, so the theories are readily grasped," said Craig Witt, also a graduate student. "If it were just to be applied cold to average convenience store customers, I don't know how it would be received."

Source: Milwaukee Journal, May 1992.

2. Investigate a single random outcome per trial.

Suppose the soft drink machine you use charges $0.75 per can. The scheme proposed by Mr. Rossides requires you to pay either $0 or $1, depending on a random number. You receive a two-digit random number between 00 and 99. If the number you select is less than 75, you pay $1. If the number you select is 75 or more, you pay nothing.

a. From a random number table, calculator, or computer, choose a random number between 00 and 99. If this represents your selection at the drink machine, how much did you pay for your drink?

b. The article suggests that things will even out in the long run. Suppose that over a period of time you purchase 60 drinks from this machine and use the random mechanism for payment each time. This can be simulated by choosing 60 random numbers between 00 and 99. Make such a selection of 60 random numbers.

i. How many times did you pay $1? What is the total amount you paid for 60 drinks?

ii. If you had paid the $0.75 for each drink, how much would you have paid for 60 drinks? Does the scheme of random payment seem fair?

c. Now, suppose you are buying a box of cookies that cost $2.43. You pay either $2 or $3, depending on the outcome of a random number selection.

i. For what values of the random number should you pay $2? For what values of the random numbers should you pay $3?

ii. Since you will pay $2 in any case, the problem can be reduced to one similar to the soft drink problem by looking only at the excess you must pay over $2. This excess amount will be either $0 or $1, just as in the case of the soft drinks. Using the rule you determined in i, simulate what will happen if you and your friends buy 100 boxes of these cookies. How many times did you have to pay the excess of $1?

iii. How much did you and your friends pay in total for the 100 boxes of cookies from the simulation? How much would you have paid for the 100 boxes if you had paid $2.43 per box? Does the randomization scheme seem fair?

3. Investigate many random outcomes per trial.

   You are taking a true/false test in history class and don't know any of the answers. You decide to guess on every question. Is this a wise decision? Let's investigate by simulation. There are five questions on the test. The simulation must be designed to give an approximate distribution for the number of correct answers on such a test.

   a. The basic random component is to make a selection of a random number correspond to guessing on a true/false question. What is the probability of guessing the correct answer on any one question? How can we define an event involving the selection of a random number that has this same probability?
   b. Each trial of the simulation represents one taking of the test; therefore, each trial must have five random selections of question outcomes within it. Select five random numbers, and with outcomes as defined in step 3a, count the number of correct answers obtained. Record this number.
   c. Repeat the procedure for a total of 50 trials (which represents taking the test 50 times). Record the number of correct answers for each trial. Construct a dot plot or stem and leaf diagram of the results.
   d. You need to get 60% correct to pass. Approximate the probability that you would get three or more correct answers by guessing.
   e. What is the average number of questions you answered correctly per trial? This is an approximation to your *expected* number of correct answers when you take the test by guessing.

Wrap-Up

1. Suppose you are now guessing your way through a 10-question true/false test. Conduct a simulation for approximating the distribution of the number of correct answers.
   a. Does the basic probability per selection of a random number change over what it was in step 3a?
   b. Does the number of random selections per trial change? If so, what is it now?
   c. Conduct at least 50 trials and record the number of correct answers for each. Make a plot of the results.

d. Again, 60% is a passing grade. What is the approximate probability of getting six or more questions correct? How does this compare with the answer to step 3d? If you had to guess at the answers on a true/false test, would you want to take a long test or a short one? Explain.
e. What is your expected number of correct answers when guessing on this test?

2. Suppose you are now guessing your way through a 10-question multiple-choice test, where each question has four plausible choices, only one of which is correct. Conduct a simulation for approximating the distribution of the number of correct answers.
   a. Does the basic probability per selection of a random number change over what it was for the true/false tests? If so, what is this probability now?
   b. What is the number of random selections per trial?
   c. Conduct at least 50 trials and record the number of correct answers for each. Make a plot of the results.
   d. What is the approximate probability of getting more than half the answers correct? How does this compare with the answer to step 1d here? Explain any differences you might see.
   e. What is the average number of correct answers per trial? How many answers would you expect to get correct when guessing your way through this test?

Extensions

1. Pairs of people eating lunch together enter a cafeteria that has square tables, with one chair on each side. Each pair chooses a separate table at which they may sit next to each other or across from each other. Suppose 100 such pairs enter the cafeteria today. Assuming that they choose seats randomly, construct an approximate distribution for the number of pairs that sit next to each other. How many of these pairs would you expect to be sitting across from each other?

2. For each of the simulations run in this lesson, make the following identifications:

   n = The number of selections per trial

   p = The probability of obtaining the outcome of interest (a "success") on any one selection

   m = The mean number of successes per trial across the simulation

   SD = The standard deviation of the number of successes per trial across the simulation

a. Do you see any relationship between m, calculated from the simulation data, and n and p? That is, can you write the expected value for the number of successes in n selections as a function of n and p?

b. Theory suggests that $(SD)^2 = np(1 - p)$. Does this rule seem to hold for your simulation data?

Assessment Questions

1. The star free-throw shooter on the girls' basketball team makes 80% of her free throws. She gets about 10 such shots per game.
   a. Set up and conduct a simulation that shows the approximate distribution of the number of successful free throws per game for this player.
   b. What is the approximate probability that she makes more than 80% of her free throws in any one game?
   c. How many free throws should she expect to make in a typical game?
   d. Over the course of a 15-game season, how many points should this player expect to have from free throws?
   e. Are there any assumptions built into the simulation that might not be realistic? Explain.

2. About 33% of the people who come into a blood bank to donate blood have type A+ blood. The blood bank under study gets about 20 donors per day.
   a. Set up and conduct a simulation that shows the approximate distribution of the number of A+ donors per day coming into the blood bank. Could you conveniently use some device other than random numbers here?
   b. If the blood bank needs 10 A+ donors tomorrow, is it likely to get them?
   c. How many A+ donors can the blood bank expect to see each day?
   d. Are there any assumptions built into the simulation that seem unrealistic? Explain.

3. Read the following article, "Love is not blind, and study finds it touching." For the 72 blindfolded people in the forehead test, simulate the distribution of the number of correct decisions that would be made had each of them just been guessing. Where does the observed value of 58 fall on this distribution? Do you agree with the conclusion that most people were not guessing but instead could actually recognize their mates?

Love Is Not Blind, and Study Finds It Touching

Blindfolded Couples Have Feel for Relationships.

Associated Press

NEW YORK—How well do lovers know each other? A new study suggests that if blindfolded, they might recognize each other just by feeling their partners' foreheads.

And if he's a man, touching his hand might do.

Seventy-two blindfolded people in the study tried to distinguish their romantic partner from two decoys of similar age, weight and height.

The blindfolded participants stroked the back of each person's right hand in one test, and the forehead in another. Each time, they were asked to pick out the lover.

Random guessing would be right 33 percent of the time. But the blindfolded people were correct 58 percent of the time in the forehead test, and women identified their man's hand 69 percent of the time.

"I think that in real life we could probably do a whole lot better," said researcher Marsha Kaitz.

The stress of being in a laboratory experiment and the carefully matched decoys probably hindered the real-world ability of recognition by touch, she said.

"I think that probably everyone can do it," Kaitz, a psychologist at Hebrew University in Jerusalem, said in a telephone interview. Touch recognition is "just a skill that has not been tapped before," she said.

Men did not show evidence of recognizing their partner's hands. Kaitz said women probably did better because the hair on men's hands made them more distinctive.

Tiffany Field, director of the Touch Research Institute at the University of Miami School of Medicine, said the finding made sense to her. Touch is an important sense in intimate contact, she said.

The experiment involved 36 heterosexual couples in their 20s who had been in their relationship for an average of two years.

Sixteen of the couples were married, and a total of 25 were living together.

Kaitz said that since the couples were relatively new in their relationship, it is possible that the touch recognition they showed is present only in the "getting to know you" phase of a relationship. She is now studying long-married couples, she said.

Source: The Gainesville Sun, Monday, June 22, 1992.

Waiting for Sammy Sosa

The geometric, or waiting-time, distribution.

Children's cereals sometimes contain prizes. Imagine that packages of Chocolate-Coated Sugar Bombs contain one of three baseball cards: Mark McGwire, Sammy Sosa, or Barry Bonds. Sammy wanted to get a Sammy Sosa card and had to buy eight boxes until getting his desired card. Sammy feels especially unlucky.

Question

Should Sammy consider himself especially unlucky? On the average, how many boxes would a person have to buy to get the Sammy Sosa card? What assumptions would you have to make to answer this question?

Objectives

In this activity, you will become familiar with the *geometric,* or *waiting-time, distribution,* including the shape of the distribution and how to find its mean.

Prerequisites

You should know how to make a histogram.

Activity

1. You will need a die or another method of simulating an event with a probability of 1/3. Roll your die. If the side with one or two spots lands on top, this will represent the event of buying a box of Chocolate-Coated Sugar Bombs and getting a Sammy Sosa card. If one of the other sides lands on top, roll again. Count the number of rolls until you get a 1 or a 2.
   a. Make a histogram of the number of rolls it took for the students to get their first Sammy Sosa cards.
   b. Describe the shape of this distribution.
   c. What was the average number of "boxes" purchased to get a Sammy Sosa card?
   d. Estimate the chance that Sammy would have to buy eight or more boxes to get his card.
   e. What assumptions did you make in this simulation about the distribution of the prizes? Do you think they are reasonable ones?

2. In some games, a player must roll doubles before continuing, such as when in jail in Monopoly. Use a pair of dice or random numbers to simulate rolling a pair of dice. Count the number of rolls until you get doubles.
   a. Make a histogram of the number of rolls the students in your class required to roll doubles.
   b. Describe the shape of this distribution.
   c. What was the average number of rolls required?

3. In steps 1 and 2, you constructed a waiting-time distribution using simulation. Now construct a theoretical waiting-time distribution for getting a different cereal prize. Boxes of Post's Cocoa Pebbles recently contained one of four endangered animal stickers: a parrot, an African elephant, a tiger, or a crocodile. Suppose 4096 children want a sticker of a parrot.
   a. How many of them would you expect to get a parrot in the first box of Cocoa Pebbles they buy? What assumptions are you making?
   b. How many children do you expect will have to buy a second box?
   c. How many of them do you expect will get a parrot in the second box?

d. Fill in Table 1.

Number of Boxes Purchased to Get First Parrot Sticker	Number of Children	Number of Boxes Purchased to Get First Parrot Sticker	Number of Children
1		11	
2		12	
3		13	
4		14	
5		15	
6		16	
7		17	
8		18	
9		19	
10		20	

Table 1

e. Make a histogram of your theoretical waiting-time distribution.

f. The height of each bar of the histogram is what proportion of the height of the bar to its left?

g. What is the average number of boxes purchased?

Wrap-Up

1. Describe the shape of a waiting-time (geometric) distribution for a given probability p of a success on each trial. Will the first bar in a waiting-time distribution always be the highest? Why or why not? The height of each bar is what proportion of the height of the bar to its left?

2. Find an example of another real-world situation that would be modeled by a waiting-time distribution.

Extensions

1. Some evidence shows that prizes are not put randomly into boxes of cereal. Design an experiment and determine how this would affect the average number of boxes that must be purchased to get a specific prize.
2. Look at the average waiting times in "Activity" steps 1, 2, and 3. Can you find the simple formula that gives this average in terms of the probability p of getting the desired event on each trial?
3. Find a formula that gives the probability that the first 5 occurs on the nth roll of a die.

Assessment Questions

1. In the game of Monopoly, you must roll doubles to get out of jail. If you haven't rolled doubles in three tries, you must pay $50. Out of every 36 people who go to jail, how many would you expect to have to pay the $50?
2. Describe some of the characteristics of a waiting-time distribution.

The Lazy Student

What happens to the spread when you add random variables.

Imagine you are taking a statistics course and your instructor keeps making you do these activities where instead of doing nice safe problems in the book, you use materials such as dice. It's horrible! These activities make you face reality time and again.

Today, you're assigned to roll a die 20 times and add the numbers you get. When you're done, the class will calculate the mean and standard deviation of all your sums.

If you're diligent, you buckle down and do it as asked. But if you're lazy, you decide on a shortcut: You'll roll the die five times, add the numbers, and then multiply the total by 4. That ought to give about the same number as 20 rolls, and it's a lot less work.

Question

If you add two random variables, what are the mean and standard deviation of the sum?

Objectives

In this activity, you will gain some practical experience with the arithmetic of random variables. It will also give you practice calculating expected values. In addition, you may see the dreadful consequences of cutting corners in statistics class.

Prerequisites

It helps to know basic concepts in probability. You also need to know how to calculate the mean and the variance (or standard deviation).

Activity

Work in pairs or small groups. In each part of this activity, your instructor will divide you into the lazy groups and the diligent groups.

1. Predicting
   Before you complete step 2, predict what will happen. Will the lazy students get the same results as the diligent ones?
2. Rolling dice
   Roll a die 20 times and add the numbers you get. (If you're lazy, just roll five times and add; multiply the result by 4.)
3. Predicting
   Look ahead to step 4. Again, predict, as quantitatively as you can, what will be the same and what will be different about the totals from the lazy and the diligent students.
4. Using the random number table
   Diligent students randomly choose 20 two-digit numbers using the random number table and add the numbers they get.
   If you're lazy, randomly pick a single digit from the table. Use that as the tens place. Then use the next five digits as ones places (see Figure 1). This gives you 5 two-digit numbers (it's "lazy" because you don't have to keep looking for a new tens digit). Repeat that process three more times so that you have a total of 20 numbers. Add them just as the diligent students do.

```
48937 86115 90632
```

Figure 1: For example, Larry, a lazy student, chose the circled "9." He immediately gets five numbers: 93, 97, 98, 96, and 91.

Wrap-Up

1. The theoretical mean (the *expected value*) of a single die roll is 3.5. Why?
2. What is the theoretical mean of the sum of 20 die rolls? Does that match what happened?
3. The theoretical standard deviation of the sum of five die rolls is about 3.76. If you multiply that by 4 (for four sets of five), you get a value a little over 15, which should be about what the lazy students got for a standard deviation. Explain why it makes sense that if you multiply a random variable by a constant, its standard deviation gets multiplied by the same constant.

4. By the same token, the theoretical standard deviation of a single die roll is about 1.71. If you multiply it by 20, you get a value over 30. Yet the diligent students had a much smaller standard deviation. Explain why it makes sense that if you add two independent random variables, their standard deviations don't add but their means do.

Extensions

1. Calculate the theoretical standard deviation of a single die roll. (The variance is 17.5/6, and the standard deviation is the square root of that. Why is the variance 17.5/6?)
2. How would you calculate the theoretical standard deviation of the sum of five die rolls?

Assessment Question

1. The SAT math and verbal scores for students who enter 4-year colleges average about 500 points, with a standard deviation of about 80. (Math scores are a little higher. Ignore that for this problem.) Suppose you were a lazy college admissions officer and decided to save work by simply doubling each applicant's verbal score rather than adding the math to the verbal for a combined score. How would the mean and standard deviation of the resulting scores compare with those of your more diligent colleagues?

What's the Chance?

Dependent and independent trials.

In many situations, people need to know the probability that a particular event will occur. For example, a pollster may wish to find the probability that voters will favor limited terms for elected officials, or a quality control manager may be interested in the probability that a product is defective in a continuous production process. As in most real-world situations, it is impossible to poll all the people or check all the products. How can we estimate these probabilities, and how accurate can our estimates be? We will use a thumbtack as our prototype and see how we can estimate the probability that a tack lands point down.

Question

What is the probability that a tack lands point down?

Objectives

In this activity, you will estimate the probability of an event. This estimate will be the relative frequency over repeated trials. The accuracy of your estimate depends not only on the number of trials but also on how the repeated trials are designed and conducted.

Prerequisites

You need to understand basic concepts in probability.

Activity

1. Collecting the data
   a. Put a tack in a small cup or jar. Place your hand or a piece of paper over the top and shake it. Do this 10 times. Note the number of times the tack lands point down and find the relative frequency of a tack landing point down.

$$\text{Relative frequency} = \frac{\text{Number of times tack lands point down}}{\text{Number of tosses}}$$

   This is your estimate of the probability that a tack lands point down.

   b. Collect the results of the tosses from the rest of the class. You want to combine the results of the tosses one student at a time and get a sequence of relative frequencies. You will now be able to complete Table 1:

Number of Tosses	10	20	30	40	50	60	70	80	90	100
Number Landing Point Down										
Relative Frequency										

Table 1

   c. Instead of tossing one tack 10 times, put 10 tacks in the cup or jar and toss the tacks *once.* Again note the number of tacks landing point down and calculate the relative frequency that a tack lands point down in this case. You now have another way to estimate the probability using a different method.

   d. Collect the results of these tosses from the rest of the class and complete a table similar to that in step 1b.

2. Analyzing the data and drawing conclusions
   Our main objective is to find the probability that a tack lands point down. *We can define the probability of a tack landing point down as the relative frequency when a tack is tossed independently a large number of times.* So the larger the number of tosses, the better our estimate of the probability.

   a. We use Table 1 to get a sequence of relative frequencies for increasing number of tosses. Do the relative frequencies in the sequence in step 1b converge toward a single value? Can we use this value as our probability that a tack lands point down? Verify that it satisfies the definition of probability. Complete the following statement: The probability that a tack lands point down is approximately ____________.

b. Let's study the sequence in step 1d, when we tossed 10 tacks at a time. Complete the following statement using the results in step 1d: The probability that a tack lands point down is approximately __________.

Do the two sequences in step 1 converge to the same value and give us the same probability? Are the experiments in steps 1a and 1c the same? Explain why you think that the estimates of the probability of a tack landing point down should be different in the two cases.

c. Does our experiment of tossing 10 tacks simultaneously meet the conditions required to estimate probability? Which of the probability estimates is more accurate?

Wrap-Up

1. We could repeat the preceding experiment tossing 20 tacks at a time. What effect would this method have on the probability estimate?
2. Can you suggest a way to get the same results for the two methods we used in the activity?

Extensions

1. We have used the thumbtack as a prototype of a situation where the probability cannot be predicted in advance. Imagine asking a sensitive question of your class, where the response will be either "yes" or "no." (Don't actually do this!) Here are two procedures you might use:

   a. Randomly sample students one at a time and observe the relative frequency of a "yes" response as your sample size increases.

   b. Ask the same question to several groups of students who have a chance to consult each other on the answers. Get the relative frequencies.

   Which of these two methods will give you a "better" estimate of the true probability of a "yes" response? Explain.

Assessment Questions

1. A good basketball player makes about 70% of her free throws. During the course of the season, her cumulative percentage of successful free throws wanders around a bit, but ends up converging to some value close to 70% for the season as a whole. In a typical season, this player gets 100 free-throw opportunities. Suppose she stood at the free-throw line and tried all 100 of these in succession. Do you think her percentage of successes would still be around 70%? Explain your reasoning.

2. A sample of 1000 students at a university are to be surveyed in an opinion poll. One way to select the sample is to randomly choose 1000 names from a student directory and interview each person chosen. Another way is to randomly select 10 names from the student directory and then let each of these 10 find 9 others to bring along to the interview. Do you think the sample percentage of students favoring the current policy on parking would be the same for each method? Explain.

Spinning Pennies

Sampling distributions. Distribution of sample proportions where $p \neq 0.5$.

When an organization conducts an opinion poll, it reports the percentage of the people sampled who favor a particular issue, such as the percentage who favor the death penalty. If the poll were repeated many times, the resulting sample percentages—one from each poll—would form a *sampling distribution.* Sampling distributions are very important in statistics, but usually, we can only imagine what the sampling distribution looks like because the people who conduct a poll don't repeat it. (If they *do* ask the same question in another poll, it is only after time has elapsed, so people's opinions may have changed.)

Question

If you toss a penny, you have a 50/50 chance of getting heads. What happens if you spin a penny and wait for it to fall? Do you still get heads 50% of the time? Suppose you spin the penny 50 times and record the sample percentage of heads. How does this sample percentage vary from one sample of 50 spins to another?

Objectives

In this activity, you learn what a sampling distribution is. By the end of the activity, you should have developed a feel for sampling distributions in general and distributions of sample proportions in particular.

Prerequisites

You should be familiar with histograms or dot plots and with the concepts of mean and standard deviation.

Activity

Your instructor will loan you a penny; you will also need the Stirling recording sheet (page 150).

1. Place your penny on its edge, with Lincoln's profile right side up and facing you. Hold the penny lightly with one finger of one hand and flick the edge of the penny sharply with a finger of your other hand to set it spinning. Let the penny spin freely until it falls. (If it hits something while spinning, do not count that trial.)
2. When the penny is at rest, record whether it is showing heads or tails. Use the recording sheet to record, in order, the results of 50 penny-spinning trials. Record the trial as a win if the penny lands showing heads and as a loss if it lands showing tails.
3. As you do your 50 trials, you will keep track of the number of heads, y; the number of tails; and the order in which heads and tails occur.
4. Compute the sample percentage of heads for the 50 observations and call this $\hat{p}$ (pronounced "p-hat"), where $\hat{p} = y/50$. This represents the process of estimating a population proportion, such as the percentage of all persons who favor the death penalty, using a sample proportion based on a sample size of 50. (Of course, most opinion polls use sample sizes of several hundred or more, but spinning the penny more than 50 times would be excessively tiring.)
5. After you have computed your value of $\hat{p}$, go to the board and add your sample percentage to the dot plot that the instructor has started. When all class data are collected, the dot plot will resemble the sampling distribution for the sample percentage of heads in 50 penny spins.

Wrap-Up

1. Write a brief summary of what you learned in this activity about sampling distributions.
2. Suppose each student in the class conducted an opinion poll in which they asked 50 randomly chosen persons whether or not they approve of the job performance of the U.S. president. If they each reported a sample percentage, what do you expect that the graph of those percentages would look like?

Extensions

1. Rather than considering the sampling distribution of the sample *percentage*, $\hat{p}$, you could consider the sampling distribution of the *number* of heads in 50 penny spins.

2. Suppose you had spun the penny only 10 times and had obtained the sequence TTHTTTHHHT. We can break this into groups as TT H TTT HHH T and say that the sequence of 10 trials contains a run of 2 tails, then a run of 1 head, a run of 3 tails, a run of 3 heads, and a run of 1 tail. There are 5 runs here, the longest of which has length 3.

   Look at the sequence of 50 heads and tails you obtained and divide the 50 trials into runs of heads and tails. Find the length of the longest run in your 50 trials. Combine your value with those of the other students in your class in a dot plot. What does this plot tell you about the longest run in 50 trials?

Assessment Questions

1. Consider the results from the class dot plot. Suppose someone who was absent made up the activity by spinning 50 pennies. What percentage, $\hat{p}$, would you expect this person to get? What is the interval in which you expect $\hat{p}$ to fall?

2. National opinion polls are often reported in the newspaper. Suppose the Gallup organization takes a random sample of 1000 voters and asks them whether they approve of the job performance of the president. What sampling distribution is related to this process?

3. When one opinion poll of 1249 adults was reported in the newspaper, the following statement was included: "Results have a margin of sampling error of plus or minus 3 percentage points." How is this statement related to a sampling distribution?

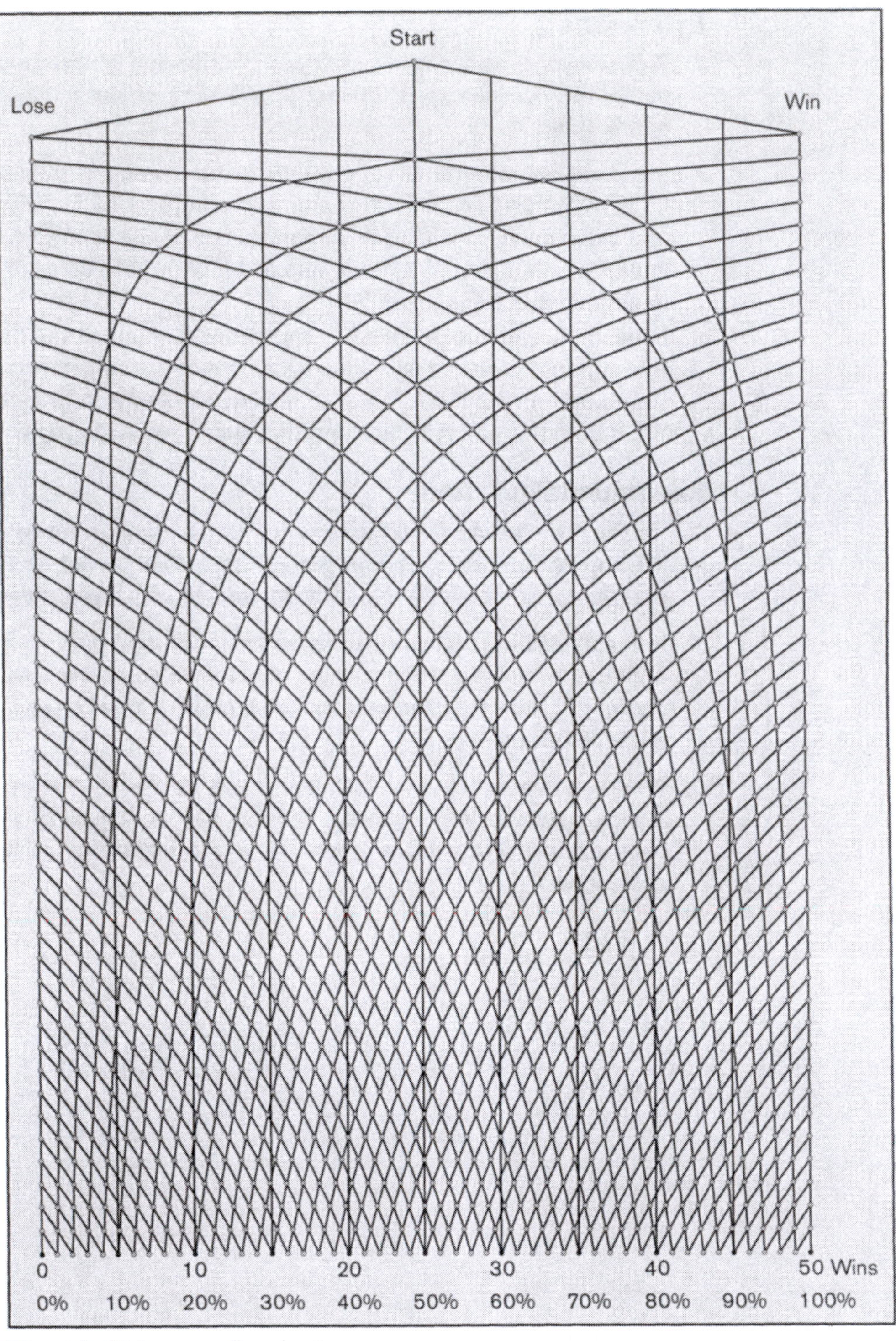

Figure 1: Stirling recording sheet

Cents and the Central Limit Theorem

How the sampling distribution of the mean of a nonnormal distribution looks normal.

Many of the variables that you have studied so far in your statistics class have had a normal distribution. You may have used a table of the normal distribution to answer questions about a randomly selected individual or a random sample taken from a normal distribution. Many distributions, however, are not normal or any other standard shape.

Question

If the shape of a distribution isn't normal, or even symmetrical, what can we infer about its mean from a random sample?

Objectives

In this activity, you will experience the consequences of the Central Limit Theorem as you observe the shape, mean, and standard deviation of the sampling distribution of the mean for random samples taken from a distribution that is decidedly not normal.

Prerequisites

You should know that the mean of a distribution is its "balance point." Also, you should understand the standard deviation as a measure of spread.

Activity

1. You should have a list of the dates from your sample of 25 pennies. Next to each date, write the *age* of the penny by subtracting the date from the current year. What do you think the shape of the distribution of all the ages of the pennies from students in your class will look like?
2. Make a histogram of the ages of all the pennies in the class.
3. Estimate the mean and the standard deviation of the distribution. Confirm these estimates by actual computation.
4. Take a random sample of size 5 from the ages of your pennies and compute the mean age of your sample. Three or four students in your class should place their sample means on a number line.
5. Do you think the mean of the values in this histogram (once it is completed) will be larger than, smaller than, or the same size as the one for the population of all pennies? Regardless of which you choose, try to make an argument to support each choice. Estimate what the standard deviation of this distribution will be.
6. Complete the histogram and determine its mean and standard deviation. Which of the three choices in step 5 appears to be correct?
7. Repeat this experiment for samples of size 10 and 25.
8. Look at the four histograms that your class has constructed. What can you say about the shape of the histogram as n increases? What can you say about the center of the histogram as n increases? What can you say about the spread of the histogram as n increases?

Wrap-Up

1. The three characteristics you examined in "Activity" step 8 (shape, center, and spread of the sampling distribution) make up the *Central Limit Theorem.* Without looking in a textbook, write a statement of what you think the Central Limit Theorem says.
2. The distributions you constructed for samples of size 1, 5, 10, and 25 are called "sampling distributions of the sample mean." Sketch the sampling distribution of the sample mean for samples of size 36.

Extensions

1. Get a copy of the current *Handbook of United States Coins: Official Blue Book of United States Coins* and graph the distribution of the number of pennies minted in each year. What are the interesting features of this distribution? Compare this distribution with the distribution of the population of ages of the pennies in the class. How are they different? Why? How can you estimate the percentage of coins from year x that are out of circulation?

2. In a geometric distribution, the height of each bar of the histogram is a fixed fraction r of the height of the bar to the left of it. Except for the first bar, is the distribution of the ages of the pennies approximately geometric? Estimate the value of r for this distribution. What does r tell you about how pennies go out of circulation? Why is the height of the first bar shorter than one would expect in a geometric distribution?

3. To cut the standard deviation of the sampling distribution of the sample mean in half, what sample size would you need?

Assessment Questions

1. What is your best guess of the mean age of all nickels in circulation? Explain.

2. Approximately how far off might this guess be? Explain. List the assumptions you are making.

Sampling Error and Estimation

How an estimate (for example, of a mean) based on a sample differs from the population value.

How far can an estimate of a parameter be from the true value of the parameter? In particular, how far can a sample mean be from a population mean? For example, all car manufacturers are required to post the average gas mileage of their cars. Typically, they take a sample of cars, drive them under different driving conditions, and calculate the average gas mileage for this sample. We assume that if everything stays the same, we can use this sample mean as an estimate of the mean of the gas mileage of all cars of that model. But how accurate is that estimate? How far is it from the true value? Can we gain some understanding of the difference between the estimate and the true value, called the *sampling error?* How often will we observe large sampling errors? If we make certain assumptions, there are theoretical results that will give us a *margin of sampling error.* For example, based on our sample standard deviation and sample size, we may be able to state that the sampling error should be between +5 and −5 in, say, 90 out of 100 cases. So we can assume that there is a 90% chance that a sample mean will be within 5 units of the unknown population mean. Another way of saying it is that the margin of sampling error is ±5 with a confidence of 90%.

Question

When you calculate the mean of a sample, what does that really tell you about the population?

Objectives

In this activity, you will learn about the sampling distribution of the sample mean, the sampling error, the margin of error, and the effect of the sample size on these quantities.

Prerequisites

You should be familiar with simple random sampling and sampling distributions. You should know how to draw random samples, compute sample means, and make stem and leaf diagrams (or draw histograms) of data.

Activity

1. Collecting the data

   Using the rectangle sheet from the earlier *Random Rectangles* activity (page 82) and a random number table, draw as many *random* samples of size 5 as required to have 100 samples for the class. Each student will calculate the average area of the sample(s) of rectangles. The instructor will collect these averages and distribute them to the class. Repeat the data collection procedure, using samples of size 10.

2. Studying sampling procedures

   a. Construct stem and leaf diagrams of the sample means for each sample size. These plots give you approximate sampling distributions of the sample means. They are approximate because they are based on only 100 samples.

   b. Guess the value of the population mean based on the stem and leaf diagrams. Show the population mean provided by your instructor on the plots. How does this value compare with your guess?

   c. Count the sample means that lie within 1 unit of the population mean. We say that these means have *sampling errors* between −1 and +1; that is, these means are at most 1 unit away from the population mean. Complete Table 1 for different values of the sampling error.

Sampling Error Between	Number of Sample Means	Proportion of Means
−1 and +1		
−2 and +2		
−3 and +3		
−4 and +4		
−5 and +5		

Table 1: Samples of size 5

   d. Repeat steps 2a–2c in Table 2 using the second sampling distribution with samples of size 10.

Sampling Error Between	Number of Sample Means	Proportion of Means
−1 and +1		
−2 and +2		
−3 and +3		
−4 and +4		
−5 and +5		

Table 2: Samples of size 10

3. Analyzing the sampling error
   a. Compare the sampling distributions for the two sample sizes with respect to the shape and spread of the distributions. Are both distributions centered at about the same point? Do the sample means have the same range of values for the two sample sizes?
   b. Using Table 1, complete the following statement for different values of the sampling error:

      A. The proportion of sample means (for sample size 5) that are within ___________ units of the population mean is ___________.
   c. What happens to this proportion of sample means as we allow the sampling error to increase; that is, what happens to the proportion as we look at different intervals of sampling errors?
   d. What happens to this proportion of sample means as we allow the sample size to increase to 10 in Table 2?
   e. The term "margin of sampling error" or "margin of error" is often used to indicate the size of the sampling error that produces a proportion of 0.95 in the answer to statement A. Find an approximate margin of sampling error for the rectangle area data with $n = 5$. Repeat for $n = 10$.

Wrap-Up

In public surveys and opinion polls, the results of the poll often give the value of an estimate as well as the margin of error for that estimate, generally associated with a confidence of 0.95. This is supposed to help readers better interpret the value of the estimate. Find a newspaper article reporting the results of a poll, including the sampling error or the margin of error. Write a paragraph explaining what the sampling error or the margin of error means in the context of the poll.

Extensions

1. If you have a choice between two estimates, the sample mean from a sample of size 5 and a sample mean from a sample of size 10, which one would generally give a "better" estimate? Clearly state your definition of "better."

2. In practice, people don't know the population mean like you did in step 2, and they don't have several estimates to get a sampling distribution of a sample mean. Suppose you have only one sample. In that case, you would need to use theoretical results. If you have studied the Central Limit Theorem, find the margins of error associated with a confidence of 0.90 and a confidence of 0.95. Compare these values with those calculated in step 2.

How Accurate Are the Polls?

How an estimate of a proportion differs from the population value. How the spread of sampling distributions defines a margin of sampling error.

A recent Gallup Poll says that 60% of the public favors stricter gun control laws and that this result was based on a survey of 1200 people. The sampling error is reported to be 3%. A pair of polls suggested that between September 20 and October 3, 2002, the percentage of likely U.S. voters favoring a ground invasion of Iraq declined from 52% to 47%, with a sampling error of 4% on each poll. How is the sampling error calculated, and what does it mean? You will investigate these questions in this activity.

Question

Why do opinion polls work, and how can their accuracy be measured?

Objectives

In this activity, you will generate *sampling distributions* of sample proportions and study their patterns for the purpose of understanding the concept of the *margin of sampling error.*

Prerequisites

You should have some experience with random sampling; some understanding of the empirical rule (95% rule) for mound-shaped, symmetric distributions; and facility at plotting distributions of data (for which a computer is helpful). Also, you should be able to calculate sample statistics, such as the sample proportion, mean, and standard deviation, and find a linear regression line.

Activity

1. Working with your team members, examine the box of beads provided by the instructor. (Your instructor may use items other than beads. For simplicity, we refer to beads throughout this activity.)
   a. Choose one color of bead, say red, which seems to occur relatively often in the box.
   b. Select a random sample of 10 beads from the box, recording the sample proportion of red beads.
   c. Repeat the sampling procedure four more times, for a total of five sample proportions for red beads. Each sample should be returned to the box before the next is selected, because all samples should come from the same population.
   d. Record the five sample proportions and give the results to the instructor.
2. Using the same bead color, repeat the procedure of step 1 for samples of size 20, 40, 80, and 100. Make sure you record five sample proportions for each sample size; give the results to your instructor.
3. Construct dot plots of the five sample proportions for each sample size on a single real number line. You may want to use different symbols for the different sample sizes. Do you see any pattern emerging?
4. Your instructor will now provide the data from all teams. Using a graphing calculator or computer, construct dot plots or histograms of the sample proportions for each sample size. (You should have five dot plots, one each for samples of size 10, 20, 40, 80, and 100.) These distributions are approximations to the sampling distributions of these sample proportions.
5. Describe the patterns displayed by the individual sampling distributions. Then describe how the pattern changes as the sample size increases. In particular, where do the dot plots appear to center? What do you think is the true proportion of red beads in the box?
6. Calculate the variance for the sample proportions recorded in each dot plot.
7. Plot the sample variances against the sample sizes, with sample size on the horizontal axis. What pattern do you see? Does it look as if the points fall on a straight line?
8. It appears that the variances are related to the sample size but the relationship is not linear. Plot the variances against (1/sample size), that is, the reciprocal of the sample size. Observe the pattern that appears. "Fit" a straight line through this pattern either by eye or by using linear regression. Does the straight line appear to fit well?

9. What is the approximate slope of the line fit through the plot of variance versus the reciprocal of the sample size? Can you see any relationship between this slope and the true proportion of red beads in the box, as estimated in step 5?

10. The preceding analysis should suggest that the slope of the line relating variance to (1/sample size) is approximately $p(1 - p)$, where p represents the true proportion of red beads in the box.
    a. Write a formula for the variance of a sample proportion as a function of p and the sample size, n.
    b. Write a formula for the standard deviation of a sample proportion as a function of p and n.

11. An interval of 2 standard deviations is called *the margin of sampling error* (or "the margin of error" or "the sampling error") by most pollsters.
    a. In repeated sampling, how often will the distance between a sample proportion and the true proportion be less than the margin of error?
    b. Using the class dot plots constructed earlier in this activity, count the number of times a sample proportion is less than 2 standard deviations from the true proportion for each plot. Do the results agree with your answer to step 11a?

Wrap-Up

1. Revisit the descriptions of poll results from the beginning of this activity. For a national Gallup Poll of adults, a typical sample size is 1500. For a poll of likely voters, the sample size is typically about 800.
    a. Are the reported margins of error correct?
    b. Interpret the results of these polls in light of the margin of error.

2. Does the concept of margin of error make sense if the data in a poll do not come from a random sample? Explain.

3. Write a brief report on what you learned about sampling distributions and margins of error for proportions.

Extensions

1. Find examples of polls published in the media. If the margin of error is given, verify that it is correct. If the margin of error is not given, calculate it. Discuss how the margin of error helps you interpret the results of the poll.

2. We developed the approximation to the margin of error in steps 10 and 11 for a single true value of p. How do we know it will work for other values of p? To see that it does, work through the steps outlined earlier for samples selected from a population with a different value of p. (The approximation does not work well for values of p very close to 0 or 1, so choose your new value of p between 0.1 and 0.9.)

Assessment Questions

1. Write a critique of a newspaper or magazine article that includes the results of a sample survey designed to estimate a proportion. The critique might be arranged according to the following outline:
   a. What are the *objectives* of the study?
   b. What is the *target population* of the study?
   c. How was the *sample* selected; was *randomization* used?
   d. What was the *method of measurement;* is *bias* a potential problem?
   e. How are the results *summarized?*
   f. Is the *data analysis* complete, or should more have been reported?
   g. What are the *conclusions* of the study?
   h. Given your knowledge of *margin of sampling error,* do you agree with the reported conclusions?
   i. If you were to do a *follow-up study,* how would you design it and what data would you collect?

   Be sure to develop the connection between margin of error and the notion of randomization in the sampling. Does it make sense to report a margin of error for a poll that was not based on a random sample?

2. Interpret sample proportions and your margins of error by relating them to decision making in the world around you. Here are two scenarios for such discussion:
   a. You have a true/false test tomorrow and don't expect to know any of the answers. That is, you will have to guess at the answers to all of the questions. You must get 70% correct to pass the test. Would you rather have a 10-question test or a 20-question test? Why?
   b. The National Football League (NFL) plays a 16-game schedule, with the best team in the league usually having around a 0.875 winning percentage and the worst team in the league having around a 0.125 winning percentage. The National Basketball Association (NBA) plays an 82-game schedule, with the best team usually having around a 0.700 winning percentage and the worst team having around a 0.300 winning percentage. Major League Baseball (MLB) plays a 162-game schedule, with the best team usually having around a 0.600 winning percentage and the worst team having a 0.400 winning percentage. It might appear that the NFL is less competitive than the NBA, which, in turn, is less competitive than MLB. Do you agree with this assessment? Why or why not?

3. The Gallup organization's Web site (http://www.gallup.com) contains a statement similar to the one presented here. After reading the statement, comment on the accuracy of their explanation of sampling error. Do you agree with this statement?

Statisticians over the years have developed quite specific ways of measuring the accuracy of samples—so long as the fundamental principle of equal probability of selection is adhered to when the sample is drawn.

For example, with a sample size of 1,000 national adults (derived using careful random selection procedures), the results are highly likely to be accurate within a margin of error of plus or minus three percentage points. Thus, if we find in a given poll that President Clinton's approval rating is 50%, the margin of error indicates that the true rating is very likely to be between 53% and 47%. It is very unlikely to be higher or lower than that.

To be more specific, the laws of probability say that if we were to conduct the same survey 100 times, asking people in each survey to rate the job Bill Clinton is doing as president, in 95 out of those 100 polls, we would find his rating to be between 47% and 53%. In only five of those surveys would we expect his rating to be higher or lower than that due to chance error.

As discussed above, if we increase the sample size to 2,000 rather than 1,000 for a Gallup poll, we would find that the results would be accurate within plus or minus 2% of the underlying population value, a gain of 1% in terms of accuracy, but with a 100% increase in the cost of conducting the survey. These are the cost value decisions which Gallup and other survey organizations make when they decide on sample sizes for their surveys.

Source: Gallup Web site (http://www.gallup.com).

IV

~

Statistical Inference

How Many Tanks?

Estimating a population from serial numbers.
Unbiased estimators.

During World War II, Allied intelligence reports on Germany's production of tanks and other war materials varied widely and were somewhat contradictory. Statisticians set to work improving the estimates. In 1943, they developed a method that used the information contained in the serial numbers stamped on captured equipment. One particularly successful venture was the estimation of the number of Mark V tanks, which have some parts whose serial numbers, they discovered, were consecutive. That is, the tanks were numbered in a manner equivalent to $1, 2, 3, \ldots, N$. Capturing a tank was like randomly drawing an integer from this sequence.

Question

How can we use a random sample of integers between 1 and N to estimate N?

Objectives

In this activity, you will develop estimators and evaluate them. Generally, there is no one answer to the question of how to estimate an unknown quantity, but some answers are better than others.

Prerequisites

You should be familiar with plots of univariate data and with summary statistics such as the mean, median, interquartile range, and standard deviation. You need not have previous experience in estimation, but you should be familiar with the importance of random sampling.

Activity

1. Collecting the data

   The chips in the bag in front of you are numbered consecutively from 1 to N, one number per chip. Working with your group, randomly select three chips, without replacement, from the bowl. Write down the numbers on the chips and return the chips to the bowl. Each group in the class should perform this process of drawing three integers from the same population of integers.

2. Producing an estimate and an estimator

   Think about how you would use the data to estimate N. Come to a consensus within the group as to how this should be done.

   Our estimate of N is ___________.

   Our rule or formula for the estimator of N based on a sample of n integers is: ____

3. Discovering the properties of the estimators

   a. The instructor will collect the data, the rules (estimators), and the estimates of N from each group and write them on a chart. Use each of the data sets (all of size $n = 3$) with your rule to produce an estimate of N. Construct a dot plot of these estimates.

   b. Calculate the mean and the standard deviation of the estimates you produced in step 3a. The instructor will place the means and standard deviations for each estimator on the chart.

   c. Collect copies of the dot plots of estimates for each of the estimators used in the class.

   d. Study the dot plots, means, and standard deviations from all estimators produced by the class. Reach a consensus on what appears to be the best estimator.

   e. The instructor will now give you the correct value of N. Did you make a good choice in step 3d? Why or why not?

Wrap-Up

1. Explain how you could use the technique developed here to estimate the number of taxis in a city, the number of tickets sold at a concert, or the number of accounts in a bank.

Extensions

1. On examining the tires of the Mark V tanks, it was discovered that each tire was stamped with the number of the mold in which it was made. A sample of 20 mold markings from one particular manufacturer had a maximum mold number of 77. Estimate the number of molds this manufacturer had used. Does this help you estimate the number of tires produced by this manufacturer? Explain.
2. Suppose the sequence of serial numbers does not begin at zero. That is, you have a sample of integers between L and N and want to estimate how many integers are in the list. How would you modify the rule used earlier to take this nonzero starting point into account?
3. Suppose the longer a tank is in battle, the greater the chance it gets captured; in addition, the longer a tank is in battle, the greater the chance it has been destroyed (and its serial number obliterated). What effects, if any, do these considerations have on your estimates of N?

Assessment Questions

1. In the context of the activity on estimating N,
   - **a.** Explain the difference between an estimator and an estimate.
   - **b.** Discuss at least two properties that a good estimator should possess.
2. You are called on to design a plan for estimating the number of people at a shopping mall on a Friday evening. Discuss how you might design this plan.
3. Suppose you were trying to estimate the number of tanks and you developed a *biased estimator*, that is, one where, on the average, your estimate would be too high or too low. Which way would you rather have it biased?

Estimating the Total of a Restaurant Bill

Sources of bias in estimation. Compensating for bias.

One way to estimate the total cost of the items on a restaurant bill is to round each of the prices to the nearest dollar before adding them. Your estimate will be too high or too low, but you would hope that in the long run the overestimates and the underestimates would more or less balance out. That is, such a method of estimation should be unbiased.

Question

If we estimate the total cost of the items on a restaurant bill by rounding the prices to the nearest dollar before adding them, will we get an unbiased estimate? If not, can we devise a method of rounding that ensures that the resulting estimate of the total is unbiased—or at least less biased?

Objectives

In this activity, you will learn the meaning of bias in estimation.

Prerequisites

You should be able to compute the mean of a probability distribution (you could also learn this concurrently). You need to use standard deviation in "Extensions" step 2.

Activity

1. **a.** Order dinner for yourself and a friend from the menu below. Your friend is paying and fantasy calories are not fattening, so don't skip dessert.
   b. Estimate the total by rounding each item to the nearest dollar and then adding the results.
   c. Find the actual total cost of the two dinners *without* rounding. Ignore taxes.
   d. Find the error in the estimate by subtracting the total cost from the estimate:

$$\text{Error} = \text{Estimated total} - \text{Actual total}$$

If the answer is negative, report it as such.

Diego and Delilah's Café

Starters, Soups, and Salads

Bruschetta	$3.49
Clam Chowder Cup	$2.95
Bowl	$4.29
Spinach, Walnut, and Feta Salad	$4.95
Nachos	$2.29

Main Courses (come with vegetables)

Bangers & Mash	$7.95
The Six-dollar Burger	$5.95
With Cheese	$6.69
Salmon with Basmati Rice	$12.79
Alamo Baby Back Ribs and Biscuits	$11.79
Roast Chicken with Apricot Sauce	$10.69
Maiale (pork loin with garlic potatoes)	$8.89
Spinach Lasagna	$8.79
Spicy Eggplant with Basil and Bean Sauce	$7.29
Feijoada (Brazilian black beans with orange)	$6.69

Dessert and Fountain Specialties

Shakes	$2.69
Malts	$3.29
Flan	$3.89
Double Chocolate Surprise	$4.95
Fresh Peach Cobbler	$4.95

Drinks

Milk	$0.95
Lemonade	$1.29
Iced Tea	$1.69
Smoothie-of-the-Day	$2.69
Sodas	$0.95

2. Make a histogram of the errors for all of the students in your class.
3. Compute the mean error for the dinners in your class. Is the rounding procedure biased? If so, in what direction? Can you suggest a reason for this?
4. Using the menu (and assuming that people order randomly from the menu), complete this probability distribution of the errors when prices are rounded to the nearest dollar.

Price Ending In	Frequency	Error When Rounded	Probability
0.29	5	−0.29	5/25
0.39	0	−0.39	0
0.49			
0.69			
0.79			
0.89	2	+0.11	2/25
0.95			

Table 1

5. **a.** Find the mean of this probability distribution using the following formula:

$$\mu_x = \Sigma(x \cdot P(x))$$

   b. What would the mean be if the method of estimation was unbiased?

6. If you were to buy one of each item on the menu and estimate the total by rounding each price to the nearest dollar, how far off would your estimate be? Do this problem two different ways.
7. Devise a method of rounding prices so that the estimate of the total bill is unbiased, or as close to it as possible.

Wrap-Up

1. Suppose you buy 10 items selected at random from the menu. If you estimate the total cost by rounding each price to the nearest dollar, how far off from the actual total would you expect to be?
2. Explain why it is impossible to underestimate the total of a bill by rounding the price of each item to the nearest 10 cents (instead of the nearest dollar).

Extensions

1. Design an experiment to determine whether the absolute value of the error in estimating the total of a bill (not necessarily at our restaurant) tends to increase, decrease, or remain the same as the number of items purchased increases. Does it matter if the rounding procedure is biased or not?

2. Use a simulation to estimate the probability that the error will be less than $1 on a purchase of 15 items selected randomly at the restaurant.

3. Collect some register tapes from grocery store purchases. Make a probability distribution of the last digits of the prices. Can you devise a method of rounding grocery store prices so that the estimate of the total bill is unbiased?

4. One way to get the distribution of final digits is to collect register tapes as in "Extensions" step 3. Another way is to look at the menu, as in the main activity about the restaurant. Describe the advantages and disadvantages of each method. Which do you think will help you develop the least biased method for estimating totals by rounding?

Assessment Questions

1. Table 2 lists the frequency distribution of the last digits of the prices of items in a grocery store. You are using the rule of rounding prices that end in 0, 1, 2, 3, or 4 down and the rest up to estimate the total cost of the items you are purchasing.

Last Digit in Price of Item	Frequency (Number of Items)	Error When Rounding	Probability
0	42		
1	13		
2	26		
3	11		
4	6		
5	50		
6	32		
7	56		
8	15		
9	28		

Table 2

a. Complete the remaining two columns of Table 2.
b. What is the average error per item with this rounding rule?
c. Suppose you buy 50 items. On average, what would be the difference between your estimated grocery bill and the estimated bill? Be sure to specify whether the estimated total tends to be too high or too low.

2. Explain what the term "biased" means.

What Is a Confidence Interval Anyway?

Explaining the confidence interval as the range of plausible population values.

Forty people are selected at random and given a test to identify their dominant eye. The person holds an 8.5 × 11 piece of paper with about a 1 × 1 inch square cut in the middle at arm's length with both hands. The person looks through the square at a relatively small object across the room. The person then closes one eye. If he or she can still see the object, the open eye is the dominant eye. If not, the closed eye is the dominant eye.

Question

Is this sample of only 40 people large enough for us to come to any conclusion about what percentage of people have a dominant right eye?

Objectives

In this activity, you will learn how to construct confidence intervals using simulation. You will also learn how to interpret confidence intervals.

Prerequisites

You should know how to simulate samples from a given binomial population by using a random number table.

Activity

1. Conduct the dominant-eye experiment just described with 40 students from your class, adding other people as necessary to bring the total up to 40. What proportion of your sample was right-eye dominant?

2. In this activity, you will be taking samples from a *population* in which 30% have some characteristic in order to see how close the proportions in the *samples* tend to come to 30%.

   a. Use a random number table to simulate taking a sample of size 40 from a population with 30% "successes." Let the digits 0, 1, and 2 represent a success and 3, 4, 5, 6, 7, 8, and 9 represent a failure. Place a tally mark in a table like the one shown here to represent your result.

Number of Successes	Frequency	Number of Successes	Frequency
0		21	
1		22	
2		23	
3		24	
4		25	
5		26	
6		27	
7		28	
8		29	
9		30	
10		31	
11		32	
12		33	
13		34	
14		35	
15		36	
16		37	
17		38	
18		39	
19		40	
20		Total	100

Table 1

b. Combine your results with other members of your class, repeating the simulation until your class has placed tally marks from 100 different samples in the frequency column of Table 1.

c. Comparing your proportion from step 1 with the frequency table from step 2b, is it plausible (that is, is there a reasonable chance) that a sample of 40 drawn from a population where 30% are right-eye dominant would have the number of right-eyed people we see in our sample? Explain.

d. Complete the following sentences based on your frequency table from step 2b: Less than 5% of the time, there were ____________ successes or fewer. Less than 5% of the time, there were ____________ successes or more.

e. In Figure 1, draw a thin horizontal box aligned with the 30% "Percentage of Successes in Population" on each side. Using "Number of Successes in Sample" on the bottom as a guide, the box should stretch in length between your two answers to step 2d.

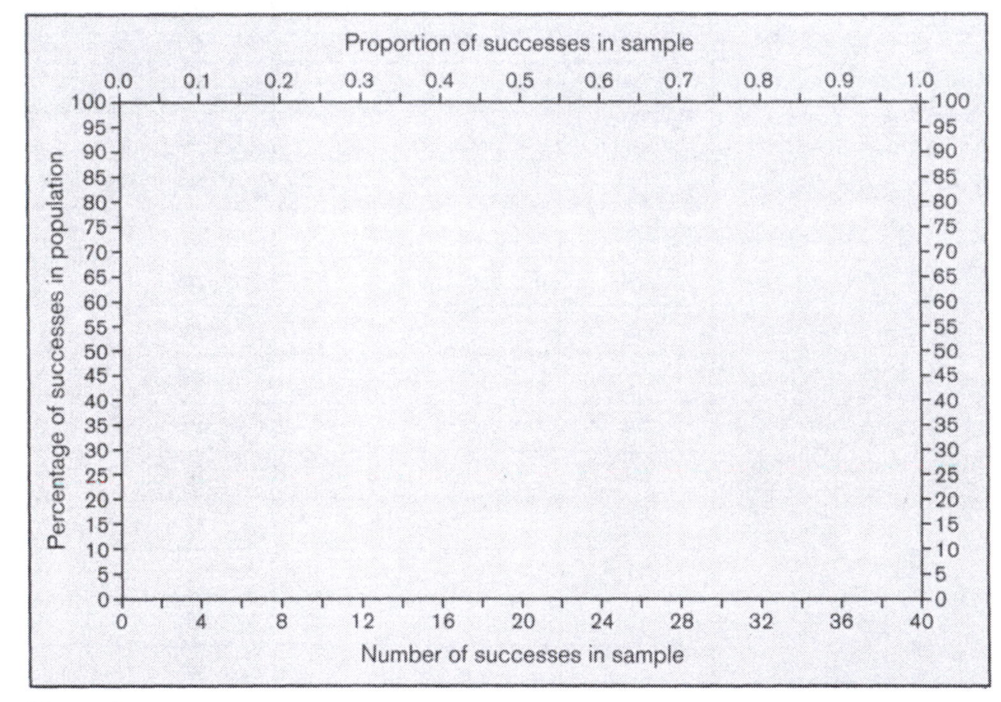

Figure 1

f. About 30% of Americans aged 19 to 28 claim that they have used an illicit drug other than marijuana. If a random sample of 40 Americans aged 19 to 28 finds 21 who claim to have used an illicit drug other than marijuana, would you be surprised? Explain.

g. According to the U.S. Bureau of Labor Statistics, about 30% of women with children younger than 6 years of age do not participate in the labor force. Would it be plausible for a survey of 40 randomly chosen mothers of children younger than 6 to find that 8 are not working? Explain.

3. The class should now divide into groups of about five students each. Your instructor will give each group one of the percentages on the sides of Figure 1. With your group, repeat steps 2a, 2b, 2d, and 2e for your new percentage.

4. Use the completed chart to answer these questions.
   a. According to the 1990 U.S. Census, about 30% of people aged 25 to 44 live alone. In a random sample of 40 people aged 25 to 44, would it be plausible to get 20 who live alone?
   b. We select a random sample of size 40 and get a sample proportion of 0.90 successes. Is this sample proportion plausible if the population has 75% successes?
   c. According to the 1990 U.S. Census, about 80% of men aged 20 to 24 have never been married. In a random sample of 40 men aged 20 to 24, how many unmarried men is it plausible to get? What proportion of unmarried men is it plausible to get?
   d. Suppose you flip a coin 40 times; how many heads is it plausible for you to get?
   e. In the 1992 presidential election, Bill Clinton got 43% of the vote. In a random sample of 40 voters, what is the largest proportion of people who voted for Clinton that it is plausible to get? The smallest?

5. Based on your sample and on the chart your class has completed, is it plausible that 60% is the percentage of the population is right-eye dominant? What percentages are plausible? These percentages are called the *90% confidence interval.*

6. Use your completed chart to find these confidence intervals.
   a. Suppose that a random sample of 40 toddlers finds that 34 know what color Barney is. What is the 90% confidence interval for the percentage of toddlers who know what color Barney is?
   b. Suppose that a random sample of 40 adults finds that 10 know what color Barney is. What is the 90% confidence interval for the percentage of adults who know what color Barney is?
   c. Observe 40 students on your campus. Find the 90% confidence interval for the percentage of students who carry backpacks.

Wrap-Up

1. Polls usually report a margin of error. Suppose a poll of 40 randomly selected statistics majors finds that 20 are female. The poll reports that 50% of statistics majors are female, with a margin of error of 10%. Use your completed chart to explain where the 10% came from.

2. People often complain that election polls cannot be right because they personally were not asked how they were going to vote. Write an explanation to such a person about how polls can get a good idea of how the entire population will vote by asking a relatively small number of voters.

Extensions

1. This activity used 90% confidence intervals because it is easy computationally to find the bottom 5% and the top 5% of a distribution. Usually, 95% confidence intervals are reported. Will 95% confidence intervals be longer or shorter than 90% confidence intervals? Explain.
2. Will the confidence intervals for samples of size 80 be longer or shorter than those for samples of size 40? Design and carry out the simulation needed to answer this question.

Assessment Questions

1. A random sample of 40 statistics students at a large university finds that 27 think the textbook is inscrutable. If you were to ask all statistics students at that university if they think their textbook is inscrutable, explain what you would expect to find.
2. You poll a random sample of 40 people and find that none of them have been drivers in car crashes that totaled the car. A fellow statistics student says, "Strange, since the sample proportion is zero, that means that the confidence interval—which gets narrow near 0 and 1—has zero width. It extends from 0 to 0." Use your understanding of confidence intervals from this activity to explain why your fellow student is wrong—and find the 90% confidence interval for this situation.
3. The most difficult idea you will probably encounter in an introductory statistics course is that of a confidence interval. What makes this idea difficult to understand?

Confidence Intervals for the Proportion of Even Digits

The meaning of the confidence level. What affects whether 95% of the confidence intervals contain the true value.

Suppose a political poll says that 56% of voters approve of the job the president is doing and that this poll has a margin of error of 3%. The 3% margin of error results from the fact that the poll was taken from a sample of voters. Because not all voters were included, there is some error due to sampling, or *sampling error.* If all voters had been asked, the polling organization predicts that the percentage would have been in the *confidence interval* of 53% to 59%. For every 100 polls that report a 95% confidence interval, the polling organization expects that 95 of the confidence intervals will contain the true population percentage. In this activity, you will take a "poll" of random digits to estimate the proportion that are even.

Question

What percentage of 95% confidence intervals will contain the true proportion of random digits that are even?

Objectives

You have learned that we expect that 95 out of every 100 of our 95% confidence intervals will contain the true population proportion. In this activity, you will test this statement by constructing confidence intervals for the percentage of random digits that are even.

Prerequisites

You should be able to compute a 95% confidence interval for a proportion using the following formula:

$$\hat{p} \pm 1.96\sqrt{\frac{\hat{p}(1-\hat{p})}{n}}$$

Activity

1. Your instructor will assign you a group of 200 random digits. Don't look at it yet.
2. When your instructor tells you to begin, count the number of even digits in your sample of 200.
3. Use the following formula to construct a 95% confidence interval for the proportion of random digits that are even:

$$\hat{p} \pm 1.96\sqrt{\frac{\hat{p}(1-\hat{p})}{n}}$$

   where $\hat{p}$ is your sample proportion, that is, the proportion of even digits you counted in your sample of 200.
4. Place your confidence interval on the class chart.
5. What is the true proportion of all random digits that are even?
6. What percentage of the confidence intervals included this proportion? Is this what you expected? Explain.

Wrap-Up

1. Bias is defined as systematic error in carrying out a survey that results in the sample proportion's tending to be too big (or too small). What were the sources of bias in your class's first survey of random digits? In which direction were the results biased?
2. Devise a method of counting the even digits accurately. Recount the number of even digits in your sample of 200 random digits and compute a new 95% confidence interval. Put these confidence intervals on a second chart. How many of the confidence intervals now contain 50%? Is this about what you expect? Explain.

Extensions

1. Investigate the kinds of systematic error that might result in bias in a political poll.
2. Investigate how major polling organizations try to minimize bias in political polls.

Assessment Questions

1. When the results of political polls are reported in the newspaper, sometimes a margin of error is given. Exactly what kinds of "error" are included in this margin of error?

2. Explain under what conditions this statement is true: Ninety-five percent of the time, the true population proportion will be in the 95% confidence interval given in political polls.

0.0	0.1	0.2	0.3	0.4	0.5	0.6	0.7	0.8	0.9	1.0

Figure 1: 95% confidence intervals for the proportion of even digits

Capture/Recapture

Estimating population size using a capture/recapture technique.

Naturalists often want estimates of population sizes that are difficult to measure directly. The capture/recapture method lets you estimate, for example, the number of fish in a lake. This idea is also the basis of methods that the U.S. Census Bureau has developed, which could be used to adjust population figures obtained in the decennial census (although using sampling techniques to adjust the census is controversial).

Question

How can we estimate the size of a population if we cannot count it directly?

Objectives

In this activity, you will learn about statistical models and see how assumptions affect statistical analyses. You will also learn about the capture/recapture method and how to use it.

Prerequisites

You should be familiar with sampling and sample proportions. For one extension, it helps if you have experience calculating the confidence interval of a proportion.

Activity

How would you estimate the number of fish in a lake or the number of bald eagles in the United States? Trying to count every fish or eagle won't work. We will demonstrate how the capture/recapture method works by trying to determine the number of goldfish in a bag of Pepperidge Farm Goldfish crackers.

1. Open one of the bags of goldfish and pour the goldfish into the "lake."
2. A volunteer should take a large sample from the lake. It's a good idea to use at least 40 or so goldfish in this sample, which often requires taking two handfuls. Someone then needs to count the number, M, of goldfish in the sample; these are the "captured" goldfish. Be sure to remember the value of M.
3. The next step is to put tags on the goldfish. If these were real fish, you could mark them with physical tags, but with cracker goldfish we mark them by "making them change color." Set aside the captured goldfish and open the second bag. For each goldfish in the captured set, put a goldfish of the other (second) flavor in the lake. (The new goldfish replace the old ones; don't put the old ones back into the lake.)
4. Shake the lake for a while to mix the two flavors of fish. Then a volunteer (it need not be the same person who took the first sample) should take a new sample of goldfish; let n denote this sample size, which should again be reasonably large. Count the number in the second sample that are tagged (call this R) and the number that are not tagged ($n - R$).
5. Now consider the percentage of fish in the second sample that are tagged (R/n). A reasonable idea is to set this percentage equal to the population percentage of tagged fish at the time of the second sample, M/N, where N denotes the unknown population size:

$$\frac{R}{n} = \frac{M}{N}$$

Solving for N, the estimate for the population size is

$$N = \frac{Mn}{R}$$

Another way to view this is to think of a two-way table that gives a cross-classification according to whether each member of the population was captured in the first phase and in the second phase.

		In First Capture?		
		Yes (tagged)	No (not tagged)	Total
In Second Capture?	Yes	R	$n - R$	n
	No	$M - R$		
	Total	M	$N - M$	N

Table 1

Discussion: Solving the equation in step 5 is easy, but the solution makes sense as an estimate of the population size only if we can say that the fraction of tagged fish we recapture is equal to the number of fish we tagged divided by the total population in the lake. This depends on many assumptions—for example, that the tags do not fall off. What else might go wrong? How would each problem affect the resulting estimate of N? Make a list of assumptions that might be violated and how the estimate of N would be affected in each case.

Wrap-Up

Suppose you wanted to estimate the number of bald eagles in the United States. Explain how you could use the capture/recapture method to do this. What assumptions might be violated? How would this affect your estimate?

Extensions

1. We have now used capture/recapture once and we have an estimate of N. However, this estimate is subject to the uncertainty that arises from the process of taking random samples (rather than counting the entire population). If we repeat the process several times, we will get different estimates of N each time. Repeat the process at least once to see how the estimate of N varies.

2. We can construct a confidence interval for the population size based on a confidence interval for the population percentage tagged. Let p denote the proportion of tagged animals in the population after the first sample (after step 4); that is, $p = M/N$. Let $\hat{p} = R/n$, the sample percentage tagged. A 95% confidence interval for p is given by $\hat{p} \pm 1.96\sqrt{\hat{p}(1 - \hat{p})/n}$. (Note that it is n, not N, in the denominator. Also, we are assuming that n is small relative to N, so we can ignore the finite population correction factor.) Setting M/N equal to the lower limit of the confidence interval for p and solving for N gives the upper limit of a confidence interval for N. Similarly, setting M/N equal to the upper limit of the confidence interval for p and solving for N gives the lower limit of a confidence interval for N.

3. The U.S. Census Bureau uses the capture/recapture idea to make adjustments to the decennial census. The capture phase is the census (persons are "tagged" by being recorded in the U.S. Census Bureau computer), and the recapture phase consists of the Post Enumerative Survey (P.E.S.) conducted after the census. In this setting the appropriate two-way table would look like Table 2.

"Captured" in the P.E.S.?	Recorded in the Official Census?		
	Yes	No	Total
Yes	R	$n - R$	n
No	$M - R$		
Total	M	$N - M$	N

Table 2

Only about 96% of all persons in the United States are "captured" in the official census. The percentage of persons in the P.E.S. who were missed in the census, around 4% or so, can be used to adjust original census figures, although whether to use the adjusted figures is a hotly debated political issue. For more discussion, see the series of articles that Stephen Fienberg has written for *Chance* magazine. (Note: The actual adjustment procedure proposed for the census is based on the capture/recapture idea but involves smoothing of estimates and becomes rather complicated.)

4. Another option is to sample fish one at a time during the recapture phase, stopping when the number of tagged fish, R, is 15, say. Thus, n, the size of the second sample, is random. This sampling method eliminates the possibility that R will be very small by chance, which would lead to a very large estimate of N in "Activity" step 5.

Technology Extensions

We could use a computer to simulate the capture/recapture process. Suppose there are actually 400 animals in the population and the first sample captures 50 of them. Then we begin the second capture phase with a population of 50 tagged animals and 350 untagged animals. That is, in your software, construct a variable—perhaps named **fish**—where the first 50 have the value **tagged** and the last 350 have the value **untagged**.

Fish	Value
47	tagged
48	tagged
49	tagged
50	tagged
51	untagged
52	untagged
53	untagged
54	untagged

Table 3: Data set with 50 tagged and 350 untagged

Learn how to sample *without replacement* using your software. If the sample size for the recapture phase is $n = 40$, then we need to simulate drawing a sample of size 40 without replacement from that population.

Equivalently, learn how to "scramble" the values; then a recapture looks at the first 40 values.

Measure	Value	Formula
R	6	count (fish = "tagged")
N	333.333	$\frac{40 \cdot 50}{R}$

Table 4

Either way, calculate R, the number of tagged fish in the sample. Then calculate the estimate for N, the population, using the equation earlier in this activity.

Finally, use the software to repeat this process many times so that you can see the distribution of estimates that you get.

Table 3 shows a few of the 400 values. Table 4 shows the calculation of the estimate; this time we had 6 tagged fish in our sample of 40, giving us an estimate of 333 fish for the population N.

Figure 1 shows a histogram of 1000 estimates of N from this type simulation.

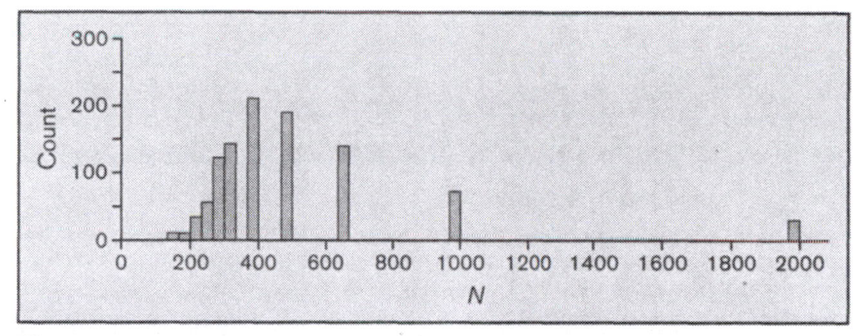

Figure 1

Note that this provides a visual display of the distribution of estimates; it does not give us a confidence interval. We could get a confidence interval, however, by using the technique in *What Is a Confidence Interval Anyway?* (page 172).

More interesting, here, is the appearance of this distribution. First, it varies a lot. It does peak at the true value, 400, but you have a good chance of being more than 20% off. Also, the simulation generates only 11 different values for N. Is that because we're simulating, or is that an inevitable consequence of the process?

Next, we can use technology to understand capture/recapture better. Rework your simulation so that the parameters are *variable,* that is, so that you can easily change the size of the population, the number of tagged fish, and the size of the recapture sample. What affects the number of different values? What affects the spread of the estimates?

Assessment Questions

1. Consider using the capture/recapture method to estimate the number of fish in a lake. Suppose 200 fish are captured and tagged in the first phase. If 348 fish are caught in the second phase and 32 of them have tags, what is the estimate of the number of fish in the lake?

2. Consider the previous question again. Suppose the tags were not attached very well and some of them fell off. How does this affect your estimate? Is it too high, too low, or about right? Why?

3. Use the data from question 1 to construct a 95% confidence interval for the number of fish in the lake.

How to Ask Sensitive Questions

Randomized response sampling. Using probability techniques to disguise survey answers and preserve confidentiality.

Pollsters often want information on issues that are considered sensitive. The IRS cannot estimate the proportion of persons who cheat on their taxes by asking a random sample of persons if they cheat, because everyone (almost) will say "no," whether it's true or not. The randomized response technique circumvents the problem of people lying when they are asked a sensitive question—it gets the needed data while protecting people's privacy.

Question

How many students in the class have ever shoplifted?

Objectives

In this activity, you will learn how to use the randomized response method. You will also review probability trees. After completing this activity, you should be able to plan and conduct a randomized response survey.

Prerequisites

You should understand how to use probability trees. You should also have had some exposure to the use of surveys, sampling, and sample proportions. In one of the "Extensions," it will also help to have learned about the confidence interval of the proportion.

Activity

Your instructor will conduct a randomized response survey of the class. In the process, you will be asked to answer a question privately. Someone will tally the class data and use the data to estimate the percentage of students in the class who have shoplifted. (You and your instructor might choose a different sensitive question to use with your class.)

You can conduct a randomized response survey in several ways. One approach involves having each student toss a coin—privately—to decide whether to answer the "real" question or a "decoy" question. You are then to tell the truth when answering whichever question is appropriate. The fact that a private coin toss has selected the question you will answer means that no one knows who is answering which question, so there is no reason to be dishonest when responding.

Although the pollster (in this case, the instructor) does not know who answered which question, it is possible to estimate θ, the population percentage who would truthfully answer "yes" to the sensitive question, by considering the probability tree shown in Figure 1.

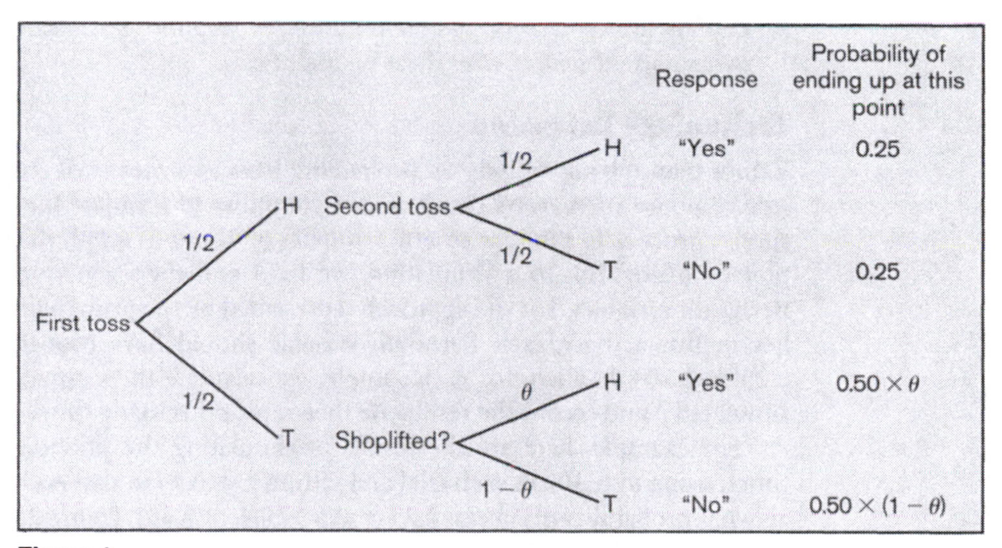

Figure 1

With this probability tree we can construct an unbiased estimate of θ. To see how, let p denote the probability that someone will answer "yes" when a pollster uses the randomized response technique. Someone can end up answering "yes" in one of two ways:

1. By getting heads twice in a row when tossing a coin
2. By getting tails on the first toss and honestly answering the sensitive question with a "yes"

The probability of getting heads twice in a row is 0.5×0.5, which is 0.25. The probability of getting tails on the first coin toss and honestly answering the sensitive question with a "yes" is $0.5 \times \theta$. Thus, $p = 0.25 + 0.5 \times \theta$.

Let $\hat{p}$ denote the proportion of "yes" responses in the sample. The expected value of the sample proportion of "yes" responses is p, which is $0.25 + 0.5 \times \theta$. We can use the sample proportion of "yes" responses, $\hat{p}$, to get an estimate of θ, which we will call $\hat{\theta}$:

$$0.25 + 0.5 \times \hat{\theta} = \hat{p}$$

Solving for $\hat{\theta}$, we have our estimate:

$$\hat{\theta} = 2 \times (\hat{p} - 0.25)$$

Wrap-Up

1. Write a brief summary of what you learned in this activity about how randomized response sampling works.
2. Explain how you could use the randomized response technique to estimate the percentage of people who cheat on their taxes.

Technology Extension

Rather than relying entirely on probability trees as a means of studying randomized response surveys, we could use the computer to simulate the randomized response process, to produce several estimates of θ, and to graph the results in a box plot or histogram. In a simulation, we fix θ and then generate a sample of n Bernoulli variables. For the approach represented by the probability tree given earlier in this activity, each Bernoulli variable should have probability of success $0.25 + 0.5 \times \theta$. Then, for each sample, we calculate the estimate of θ from the simulated $\hat{p}$ and record the result. We then repeat this many times (say, 250 times).

For example, here are the results of simulating the previous approach 250 times, using $n = 100$ in each trial and setting $\theta = 0.4$ (so that each Bernoulli variable has probability of success $0.25 + 0.5 \times 0.4$, or 0.45). Figure 2 summarizes 250 simulated randomized response surveys. Note that the randomized response technique provides estimates of θ that center at approximately the "true" value of 0.4.

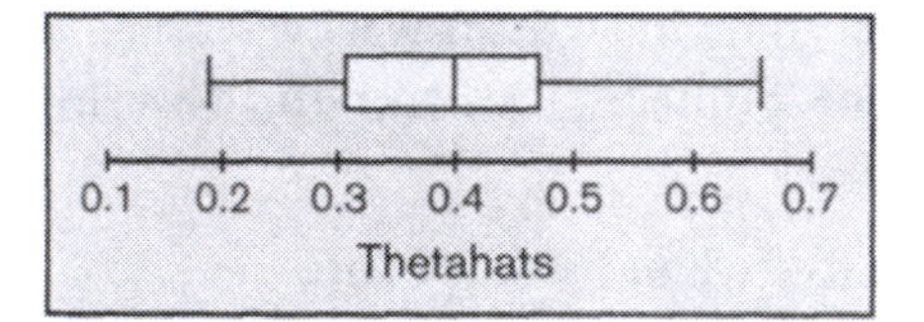

Figure 2

Options

There are other ways to do a randomized response survey that resemble the one we have already used but differ in the details. We'll describe two more methods and use simulation to see how well they do at estimating θ.

Just to recap, though, here's what we did:

Plan A (What We Already Did)

You flip two coins privately. If the first toss is heads, the second toss determines your answer: heads is "yes," tails is "no." If the first toss is tails, just answer the sensitive question, "Have you ever shoplifted?"

Plan B (A Slightly Different Technique)

Toss a coin once, privately. If the coin lands heads, your answer is "yes." If the coin lands tails, answer the question "Have you ever shoplifted?" (or whatever sensitive question you have chosen). As before, you can be reassured that you can answer honestly. Note that this is almost the same as Plan A, except that here, if the first toss is heads, you *always* answer "yes" instead of waiting for the second toss.

In this setting, the probability tree is almost the same as before, except that one branch is simpler. See Figure 3.

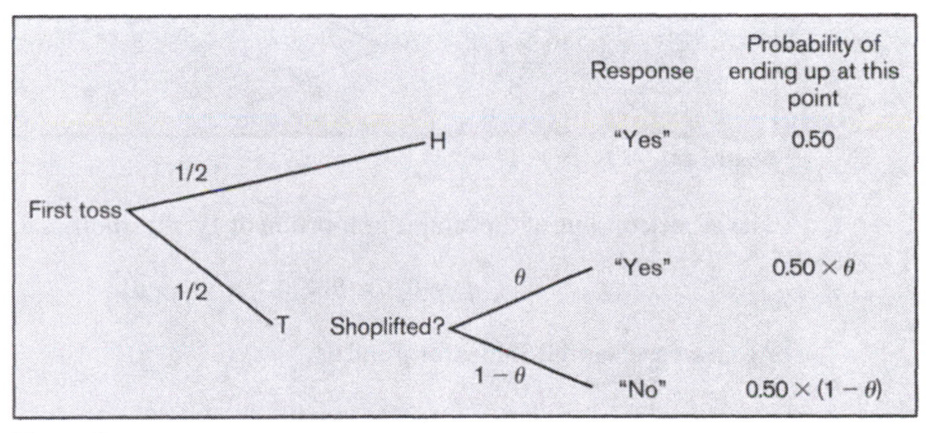

Figure 3

The expected value of the sample proportion of "yes" responses is p, which is $0.5 + 0.5 \times \theta$. This leads to the equality:

$$0.5 + 0.5 \times \hat{\theta} = \hat{p}$$

Solving for $\hat{\theta}$, we have

$$\hat{\theta} = 2 \times (\hat{p} - 0.5)$$

Unlike the case of the first randomized response method (Plan A), with this method a "no" response clearly means that the person has not shoplifted.

Plan C, Known as the Warner Approach

There are yet other ways to conduct a randomized response survey. For example, you (the respondent) choose a random digit. (One way to do this is to take out a dollar and use the last digit of the dollar's serial number.) If the digit is 4, 5, or 6, give a truthful answer to the sensitive question. If the digit is 0, 1, 2, 3, 7, 8, or 9, you should lie—that is, give the opposite of the truthful answer. So, for example, if my random digit is 7 and I have not shoplifted, I would answer "yes."

In this setting the probability tree looks like that shown in Figure 4.

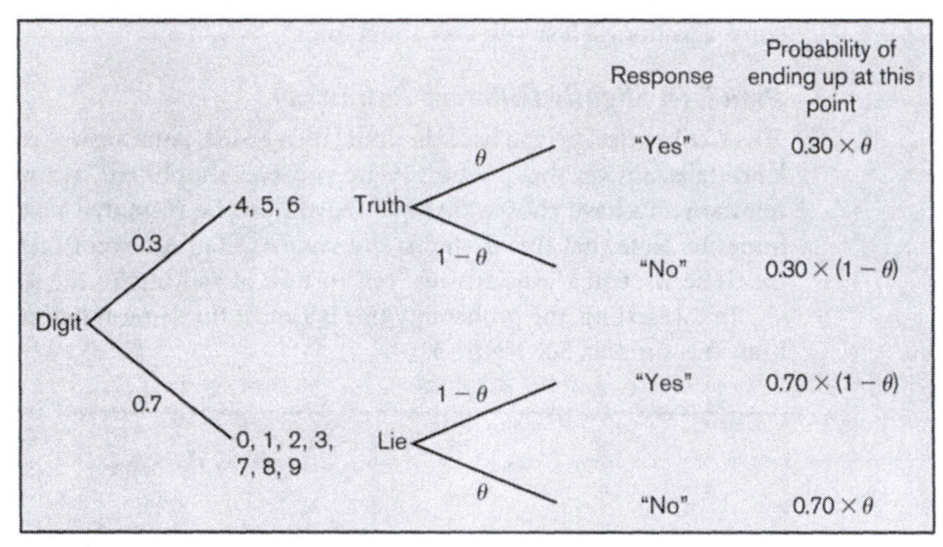

Figure 4

The expected value of the sample proportion of "yes" responses is p again, which is

$$p = 0.3 \times \theta + 0.7 \times (1 - \hat{\theta})$$

Again we use sample values for $\hat{p}$ and $\hat{\theta}$:

$$\hat{p} = 3 \times \hat{\theta} + 0.7 \times (1 - \hat{\theta})$$

Solving for $\hat{\theta}$, we have

$$\hat{\theta} = (0.7 - \hat{p})/0.4$$

Note that here the standard error of the estimate of θ is $2.5 \times \sqrt{\hat{p}(1 - \hat{p})/n}$, which differs from the standard error for the first two methods.

Technology Extensions, Continued

We can simulate each of these techniques and compare the results.

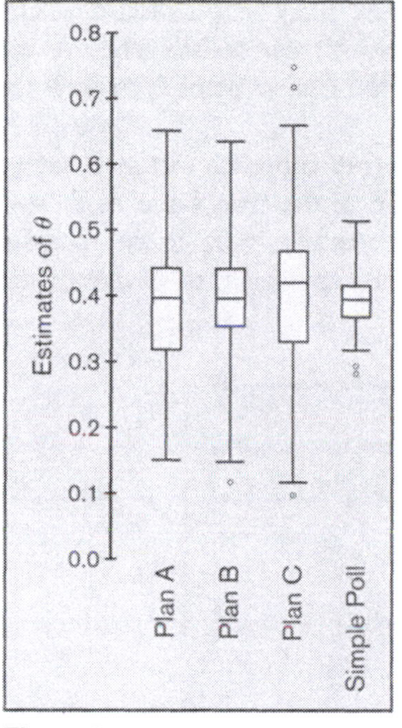

Figure 5

For example, Figure 5 contains the results of simulating the three preceding approaches 250 times each, using $n = 100$ in each trial and setting $\theta = 0.4$. The fourth box plot, labeled "Simple Poll," is the result of simulating a simple random sample, again with $n = 100$ and $\theta = 0.4$, in which the randomized response method is not used.

Each of the first three box plots summarizes 250 simulated randomized response surveys, and the fourth summarizes 250 simulated simple polls. All four provide estimates of θ that center at about 0.4, but Plan C (the Warner approach) produces estimates that are more variable than those from the first two methods.

Note as well that with $\theta = 0.4$, the expected value of $\hat{p}$ is 0.45 for Plan A but 0.7 for Plan B, so the results from Plan B are a bit less variable than the Plan A results [since $0.45 \times (1 - 0.45) > 0.7 \times (1 - 0.7)$]. The three randomized response simulations yield more variable results than those given by the simple poll simulation. Remember, using the randomized response method effectively cuts the sample size in half; the higher variability of the estimates is the price one must pay to get truthful answers when asking a sensitive question.

Experimental Design

If we want to ask sensitive questions using one of these techniques, we have many decisions to make; for example, if we use the Warner approach (Plan C), how many digits should we use for the liars? How many people should we sample? We can use technology to study the consequences of these decisions before we make them. Our goal is to learn what we can about the true value of θ, so we'll want to reduce the variability in $\hat{\theta}$.

For example, if you use the Warner approach outlined here, the variability of $\hat{\theta}$ depends on the total sample size n, the true value of θ, and the probability (call it π) that a respondent tells the truth ($\pi = 0.3$ in the preceding example). You could design and conduct a factorial experiment to study how these three factors influence the variability of $\hat{\theta}$.

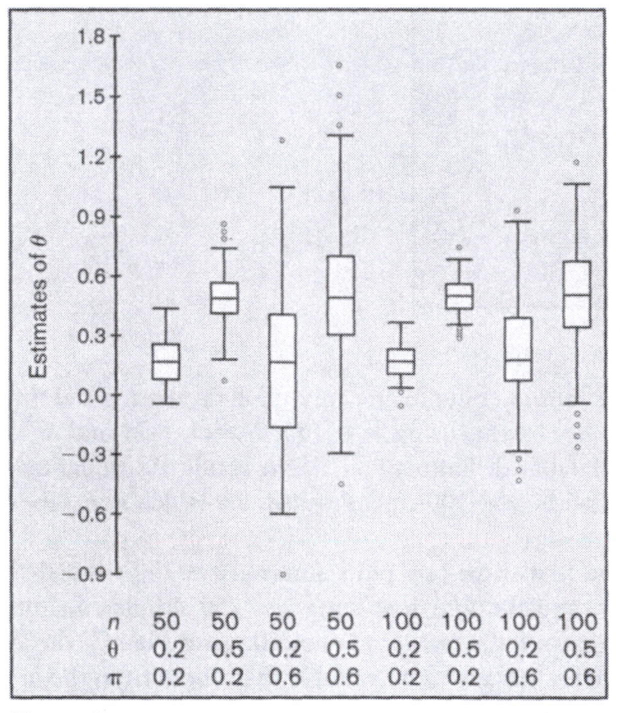

Figure 6

For example, Figure 6 shows box plots of 250 values of $\hat{\theta}$ from each of the eight combinations that arise when n is either 50 or 100, θ is either 0.2 or 0.5, and π is either 0.2 or 0.4. Note that all eight box plots are centered at θ, which shows that the process is unbiased. The effect of n is modest but noticeable, the effect of θ is small, and the effect of π is considerable. When $\pi = 0.6$, it is quite common to get an estimate of θ that is outside the range (0,1).

Assessment Questions

1. Consider using the first version (Plan A) of the randomized response technique with a group of $n = 60$ persons. Suppose that 36 persons answer "yes" and 24 answer "no." Use these data to estimate θ.

2. Is it possible to get a negative estimate of θ when using the randomized response technique? Can the estimate of θ be greater than 1? If so, how can this happen?

Estimating a Total

It is not always clear how to sample to get the best estimate.

Pine forests are periodically infected with beetles, and the infestations often seem to jump from tree to tree randomly. Diseases that spread through the soil, however, tend to infect large clumps of trees in a few locations. How can a forester choose a good sampling method to count diseased trees? Should the sampling method be different for the two types of disease?

Question

How can we construct a good estimate of the number of diseased trees in a forest by counting diseased trees in sampled plots? How should we select the size and number of the sampled plots?

Objectives

In this activity, you will learn one way to estimate a total and how systematic sampling may be used to good advantage in certain situations.

Prerequisites

You should have used the sample mean to estimate a population mean. You should also be familiar with the notion of random sampling and the concept of standard error of the mean.

Activity

Figure 1 (page 198) shows a number of dots scattered within a rectangle that measures approximately 18 cm × 10 cm. The goal is to estimate the number of dots in the rectangle. (The dots may represent diseased trees in a forest.)

1. Collecting the sample data randomly
   a. To select a sample plot, randomly drop a quarter inside the rectangle. To obtain the data on a sampled plot, draw a circle around the quarter and count the dots inside or touching the circle. Repeat this procedure until you have selected a sample of six plots and collected the data. (The sampled plots may overlap. Do not use information in the pattern when selecting sample plots; if this were a real forest, you would not be able to see the pattern of diseased trees.)
   b. Calculate the mean and the standard deviation for these six counts.
   c. A quarter is approximately 4.9 cm^2 in area. Use this information, along with the data from the random sample, to estimate the total number of dots inside the rectangle of Figure 1.

2. Collecting the sample data systematically
   a. Another way to select the sample plots is to arrange the circles *systematically* across the rectangle of Figure 1. For example, you could have two rows of three evenly spaced circles each. (A chess board is a systematic arrangement of differently colored squares. Student desks in a classroom are usually arranged systematically—in rows—rather than randomly.) Sampling with a quarter, select six sample plots systematically and count the number of dots in each.
   b. Calculate the mean and the standard deviation for these six counts.
   c. Use the data from the systematic sample to estimate the total number of dots inside the rectangle of Figure 1. Compare this estimate with the one from random sampling.

3. Changing the size of the sample plot
   The size of each sample plot will now be changed to a dime instead of a quarter. A dime is approximately 2.5 cm^2 in area, about half the size of a quarter. Thus, to keep the sampling effort equivalent to that used in steps 1 and 2, 12 sample plots should be used when sampling with a dime.
   a. Follow the instructions of step 1 with 12 dime-sized sample plots.
   b. Follow the instructions of step 2 with 12 dime-sized sample plots.
   c. Do you see any advantages or disadvantages of using 12 dime-sized plots rather than 6 quarter-sized plots for estimating the total number of dots?

4. Combining the data for the class

   The instructor will now provide you with the estimates of the totals from each member (or team) in the class. Keep in mind that we are using four different estimators: quarter-random, quarter-systematic, dime-random, and dime-systematic.

   a. Construct four plots, one for each of the four sets of estimates. (Dot plots or stem and leaf diagrams work well.)
   b. Compare the four plots. How do the four methods compare? Does one method look like it is clearly better than the others?
   c. Where do the plots appear to center? How many dots do you think are in Figure 1?

5. Figure 2 (page 198) contains another array of dots. Repeat steps 1 through 4 to compare estimates of the total number of dots in Figure 2. Is there a preferred method among the four for estimating the total in this case? (The rectangle of Figure 2 is approximately 18 cm × 10 cm.)

Wrap-Up

1. You are to estimate the total number of TV sets in your city or county.
   a. Explain how you would design the sampling plan to collect the data.
   b. Explain how you would estimate the total number of TV sets from the data you would collect.

Extensions

1. You may recall from previous studies that the standard deviation of a sample mean (or the standard error of the mean) is estimated by dividing the standard deviation of the sample data by the square root of the sample size. This estimate of error can be used to form an estimate of error appropriate for the estimation of a total.
   a. The estimates of the total number of dots were of the form $K \times$ (sample mean), where K is some known numerical constant—in this case, the ratio of the area of the rectangle to the area of the coin. Produce a formula for the standard error of an estimate of this form.
   b. Using the appropriate K, calculate an approximate standard error for the estimates of the totals constructed earlier in this activity. Use only your own sample data in this calculation. (Recall that there were eight estimates.)
   c. Do these standard errors help in deciding how the methods compare? If so, do there appear to be major differences among the four methods for Figure 1? How about for Figure 2?

Assessment Questions

1. You are to estimate the number of cars per hour that flow into a mall parking lot over a weekend. You have 4 hours of time in which to observe the lot and collect the data. Should you observe the lot for many small time periods over the weekend or for a few large ones? Explain.

2. You are to estimate the total number of employees in local businesses by taking a sample of businesses from the yellow pages of the local telephone directory.
   - **a.** Explain how you would select the businesses that make up the sample.
   - **b.** Assuming you can contact the sampled businesses and determine how many employees they have, explain how you would construct an estimate of the total number of employees.
   - **c.** Do you think estimates conducted in this way might tend to be too low? Explain.
   - **d.** Explain how you would attach a measure of error to the estimate found in question 2b.

3. You are to estimate the total number of trees growing on public property in a moderately large city. You are provided with a map of the city that highlights the public property. You must decide whether to sample a few large parcels of property or a much larger number of small parcels of property, given that the total acreage covered would be the same in both cases. What factors would influence your decision and why?

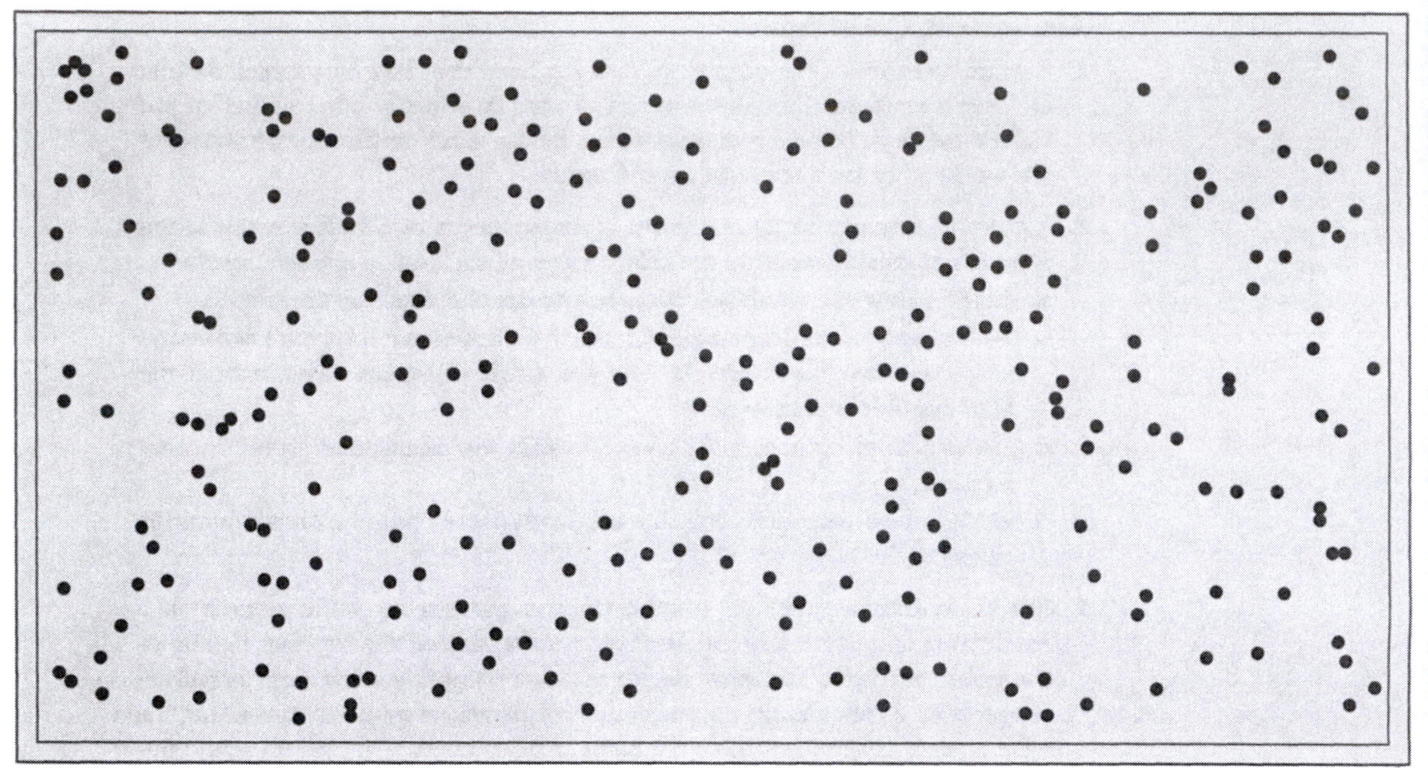

Figure 1

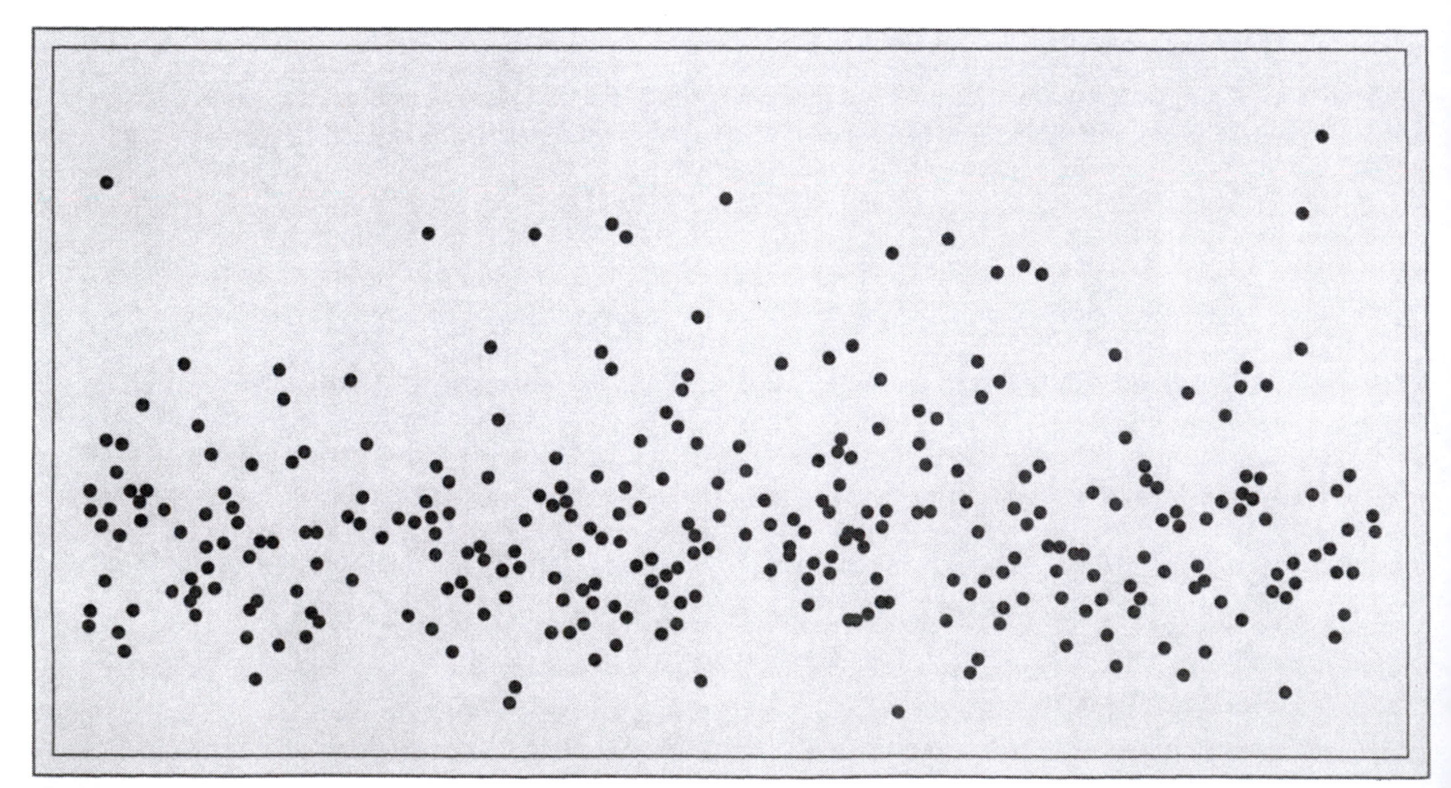

Figure 2

The Bootstrap

Creating an interval estimate for statistics when the traditional confidence interval may be inappropriate.

Annual rainfall varies from place to place and from year to year. Although there are no obvious trends in the data, Los Angeles has had both periods of near-drought conditions and also years with heavy rainfall totals. Figure 1 is a histogram of the annual rainfall totals for Los Angeles for the years 1943 to 1992.

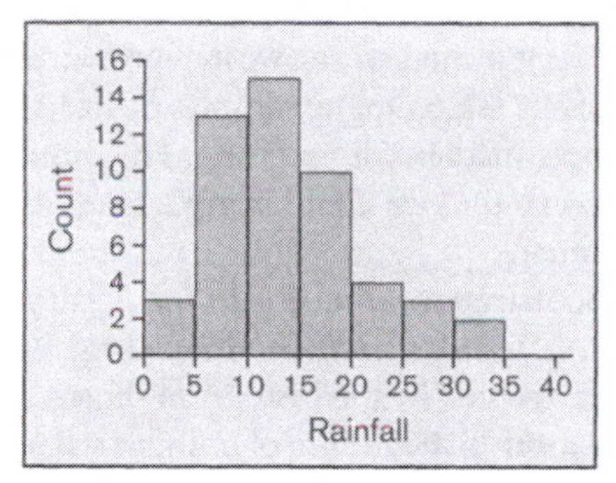

Figure 1

Note that the distribution is skewed. Suppose we consider the 50 rainfall values to be a sample from a larger population and we want to estimate the median of that population. We can use the sample median, 13.075, as a point estimate of the population median, but we do not have a handy formula to use for a confidence interval. The *bootstrap* is a general statistical technique that uses the computer for such things as constructing interval estimates of parameters.

Question

How much rain can Los Angeles expect to get in a typical year?

Objectives

In this activity, you will develop an understanding of the bootstrap technique. After completing this activity, you should understand how the bootstrap works and be able to use the bootstrap method to construct confidence intervals for parameters such as the median of a distribution.

Prerequisites

You should be familiar with confidence intervals. The more familiar and comfortable you are using the computer, the easier it will be for you to complete this activity.

Activity

Following are the rainfall totals (in inches) for the years 1943 to 1992:

22.57	17.45	12.78	16.22	4.13	7.59	10.63	7.38	14.33	24.95
4.08	13.69	11.89	13.62	13.24	17.49	6.23	9.57	5.83	15.37
12.31	7.98	26.81	12.91	23.66	7.58	26.32	16.54	9.26	6.54
17.45	16.69	10.70	11.01	14.97	30.57	17.00	26.33	10.92	14.41
34.04	8.90	8.92	18.00	9.11	11.57	4.56	6.49	15.07	22.56

We will use these data to make an inference about Los Angeles rainfall.

The sample median of 13.075 describes the data we have. If we got a new set of 50 years of rainfall data, we would expect the new sample median to differ from 13.075. The question is, "How does the sample median behave in repeated samples of size 50?"

Ideally, we would study this by taking many samples of size 50 from the population and computing the sample median for each one. This would require having the entire population before us so that we could sample from it! However, all we have are the 50 values listed earlier.

The key idea behind the bootstrap technique is to *use the distribution of the 50 sample values in place of the true population distribution.* That is, we will think of the distribution we have, as represented by the histogram shown in Figure 1, as being an estimate of the population distribution of rainfall values. We will use a model that says that in some years Los Angeles will get 22.57 inches of rain, in other years it will get 17.45 inches, and so on, and the only rainfall totals possible are the 50 values listed earlier. Of course, we would be better off if we had a sample of size 100, rather than of size 50—the larger the sample size, the better we expect the sample histogram to represent the true population distribution.

To see how medians of repeated samples of size 50 behave, we will sample *with replacement* from the set of 50 values listed earlier. Thus, in a sample of 50 years, we might get 1 year in which 22.57 inches of rain fall, 2 years in which 17.45 inches of rain fall, no years in which 12.78 inches of rain fall, and so forth.

It's a good idea to simulate this process by hand before moving to the computer. Write the rainfall totals on slips of paper and put the slips in a bag. Then draw out a slip, record the value, and replace the slip in the bag. Do this until you have a total of 50 observations. This collection of 50 draws is the *bootstrap sample.* (It is possible that you will draw each of the 50 slips exactly once, but this is highly unlikely. It is more likely that you will draw some slips several times and others not at all.) Now compute the median of your bootstrap sample. This is a bootstrap estimate of the population median.

Using the Computer

We would like to see how the sample median behaves in repeated samples of size 50. Drawing a sample by hand takes a lot of time. We'll use the computer to repeat the process very quickly, drawing 600 bootstrap samples and computing 600 bootstrap estimates of the population median.

Once you have 600 bootstrap estimates, you can use them to construct an interval estimate of the population median. (You can use the 600 values to study other aspects of the median as well.)

To get a 95% "bootstrap interval," we take the middle 95% of this distribution. We throw away the top 2.5% and the bottom 2.5% of the distribution. That is, we sort the 600 values and throw out the first 15 (2.5% of 600 is 15) and the last 15. This leaves the middle 570 values, which gives us a 95% bootstrap interval for the population median.

Using Fathom

Here are specific instructions for using Fathom to do this:

1. Store the data in a collection. Give the attribute a suitable name such as **rainfall.**
2. Select the collection and choose **Sample Cases** from the **Analyze** menu. A sample collection will appear.
3. Double-click the sample collection to open its inspector. Set it to sample with replacement (which is the default), 50 times (since there are 50 years in the original data). Turn animation *off* to save time.
4. Click on the **Measures** tab in that inspector. Make two new measures: **BootMedian** and **BootMean.** Their formulas should be **median(rainfall)** and **mean(rainfall)**, respectively. Make sure reasonable numbers show up in the **Values** column. Now close the inspector.
5. Select the sample collection and choose **Collect Measures** from the **Analyze** menu. A "Measures from Sample" collection will appear.
6. Double-click the new measures collection to open its inspector. Tell it to collect 600 measures and turn off animation for speed. Click **Collect More Measures** to make it happen.
7. Click on the **Cases** tab of the inspector. Make a new graph and drag **BootMedian** from the inspector to the horizontal axis. You'll get a dot plot. Change it to a histogram using the menu in the corner of the graph.
8. With the graph selected, choose **Plot Values** from the **Graph** menu. To see the lower end of the bootstrap interval, enter the formula **percentile(2.5,).** (The blank after the comma is correct; it tells Fathom to find that percentile of whatever is on the axis.) Click **OK** to close the formula editor and make the value appear. Do the same for the top end (use 97.5 instead of 2.5).

You have already bootstrapped the mean—just use **BootMean** instead of **BootMedian.** To bootstrap any other statistic, just make a measure for it as in step 4 and recollect measures as in step 6.

Figure 2 is a histogram of 600 bootstrapped medians generated using Fathom. From the figure, the 95% bootstrap interval for the rainfall median is (~10.9, ~15.1). Think about what this means. For example, 2.5% of the time, the bootstrap median (the median of a sample of 50 drawn with replacement) was less than 10.9 inches. Of course, repeating the bootstrap process will yield a slightly different interval each time.

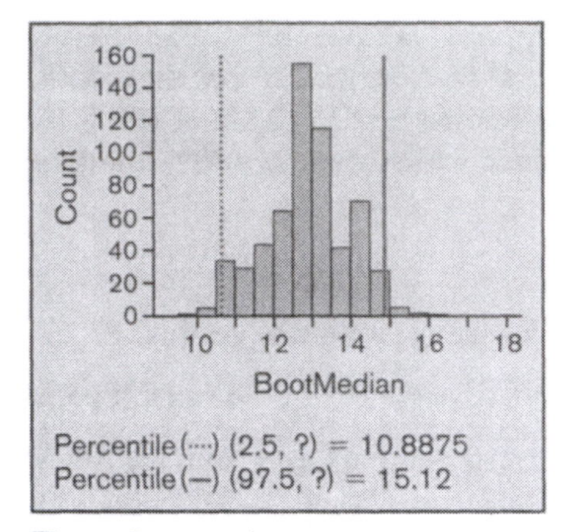

Figure 2

By taking 600 bootstrap samples we get a pretty good idea of what the distribution of all possible bootstrap samples looks like. However, this does not necessarily mean that our interval will be good. The accuracy of the bootstrap interval depends on how well the original 50 observations represent the true population—the key feature is the quality of original sample. If the original sample is biased in some way (for example, if there is an overall trend in the rainfall) or if the sample size is simply too small for the sample histogram to give a good representation of the population, then taking 6000 or even 60,000 bootstrap samples (rather than 600) won't produce a good result.

Wrap-Up

1. Write a brief summary of what you learned in this activity about the bootstrap technique and how it works.

2. Explain how the bootstrap technique could be used to construct an interval estimate of the 90th percentile of a distribution.

Extensions

You might want to verify that the bootstrap gives sensible results by using it in a situation in which normal theory works. That is, you could bootstrap the mean with data that came from a normal distribution and compare the bootstrap interval with the usual confidence interval for a population mean.

Assessment Questions

1. A sample of 13 college students were asked, "How much did you spend on your last haircut?" Here are their responses (in dollars): 0, 10, 10, 0, 12, 23, 77, 35, 17, 14, 12, 7, 12. Use these data to construct a bootstrap interval for the median number of dollars that a student at this college spends on a haircut.

2. Suppose we wanted an interval estimate of the population mean, rather than the population median. Why might we prefer using the bootstrap technique to using the standard formula $\bar{x} \pm ts/\sqrt{n}$?

Statistical Evidence of Discrimination

Using the randomization test to show that variables are associated.

In 1972, 48 male bank supervisors were each given the same personnel file and asked to judge whether the person should be promoted to a branch manager job that was described as "routine" or whether other applicants should be interviewed. The files were identical except that half of them showed that the file was that of a female and half showed that the file was that of a male. Of the 24 "male" files, 21 were recommended for promotion. Of the 24 "female" files, 14 were recommended for promotion. (From B. Rosen and T. Jerdee [1974], "Influence of sex role stereotypes on personnel decisions," *J. Applied Psychology,* 59:9–14.)

Question

Is the opening example convincing evidence that the bank supervisors discriminated against female applicants, or could the difference in the numbers recommended for promotion reasonably be attributed to chance? That is, is it possible that there was no discrimination on the basis of sex: It just happened that 21 out of the 35 bank managers who recommended promotion got files marked "male."

Objectives

In this activity, you will learn how to use the randomization test to tell if the difference between two proportions is *statistically significant.* The randomization test will be a simulation of the situation under the hypothesis that there was no discrimination on the basis of sex. "Statistically significant" means that a difference as big as or bigger than that which occurred is unlikely to have happened without a cause other than random variation. When you have finished this activity, you should be able to use the randomization test to decide whether you can reasonably attribute the difference between two proportions to chance or whether you should search for some other explanation.

Prerequisites

You should have had some previous experience with simulation.

Activity

1. Remove all four aces from a deck of playing cards. There will now be 24 red cards in the deck, which will represent "male" files, and 24 black cards in the deck, which will represent "female" files. Shuffle the cards at least seven times and then cut them.
   a. Count out the first 35 cards to represent the files recommended for promotion. How many males were in this pile?
   b. On a number line like that shown in Figure 1, place an X above the number of males you got in step 1a.

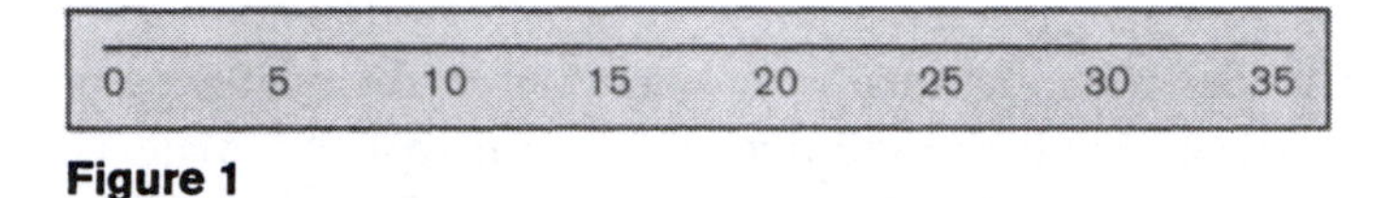

Figure 1

   c. In this simulation, what is the probability that a particular male will be promoted? A female? Does this simulation model the case of discrimination on the basis of sex or the case of no discrimination?
   d. Repeat the simulation with the cards until you have a total of 100 X's on the number line. You may want to combine results with other groups. (If you have 10 groups, each group would do only 10 simulations.)
   e. Use the results on your number line to estimate the probability that 21 or more of the 35 recommended for promotion will be male if there was no discrimination on the basis of sex.
   f. Do you believe your simulation provides evidence that the bank supervisors discriminated against females?

Few situations in real life are as clear-cut as that of the bank supervisors. In real life, the files are never identical. Education, experience, character, recommendations, and test scores vary. Statistics can tell us only whether the difference between two groups can reasonably be attributed to chance. If it cannot reasonably be attributed to chance, then further digging is required to determine whether the explanation is discrimination or a difference in qualifications.

In *Griggs v. Duke Power Company* (1971), the U.S. Supreme Court established the idea of "disparate impact." Disparate impact occurs, for example, when the pass rate of one group on an employment test is substantially lower than that of another. Such an employment test is illegal unless the employer can prove that the use of the test is a business necessity. For steps 2 and 3,

   a. Design a simulation with cards to help you determine whether the difference in pass rates can reasonably be attributed to chance or whether there is clearly disparate impact on this examination.
   b. Repeat your simulation 100 times, sharing the work with other groups. Place your results above a number line.
   c. Write a paragraph explaining your conclusions.

2. In the 1975 court case *Chicano Police Officers Association v. Stover*, 3 of 26 Chicanos passed an examination and 14 of 64 whites passed the same examination.
3. In the 1977 *Dendy v. Washington Hospital Center* case, 26 out of 26 white nurses passed an examination and 4 out of 9 African-American nurses passed the same examination.

Wrap-Up

1. Discuss why the procedure you have been using may have been named the "randomization" test.
2. For a classmate who is absent today, write an explanation of how to use the randomization test to determine whether there is discrimination in hiring.

Extensions

1. Joseph Lister (1827–1912), surgeon at the Glasgow Royal Infirmary, was one of the first to believe in Pasteur's germ theory of infection. He experimented with carbolic acid to disinfect operating rooms during amputations. When carbolic acid was used, 34 of 40 patients lived. When carbolic acid was not used, 19 of 35 lived.
   a. Design a simulation with cards to help you determine whether the difference can reasonably be attributed to chance or whether there is evidence that carbolic acid saves lives.
   b. Repeat your simulation 100 times, sharing the work with other groups. Place your results above a number line.
   c. Write a paragraph explaining your conclusions.
2. In a psychology experiment (Schachter, 1959), a group of 17 people were told that they would be subjected to some painful electric shocks and a group of 13 people were told they would be subjected to some painless electric shocks. The subjects were given the choice of waiting with others or waiting alone. Table 1 shows the results.

	Wait Together	Wait Alone
Painful Shocks	12	5
Painless Shocks	4	9

Table 1

   a. Design a simulation with cards to help you determine whether the difference in the proportions that chose to wait together can reasonably be attributed to chance.
   b. Repeat your simulation 100 times, sharing the work with other groups. Place your results above a number line.
   c. Write a paragraph explaining your conclusions.

Assessment Questions

1. Suppose that on the 1994 test for the fire department, 5 out of 10 women pass and 9 out of 10 men pass. Assume for a moment that the chance a man passes this examination is the same as that of a woman passing. Describe a way to simulate with cards the probability of getting a difference between the proportion of men who pass and the proportion of women who pass that is as large or larger than it was in 1994.

2. Suppose that on a hiring examination, 35 out of 40 women pass and 20 out of 40 men pass. Under the assumption that men and women are equally likely to pass, a simulation was performed 500 times. Table 2 shows the number of men who passed. Write a paragraph explaining what conclusion should be drawn from this simulation.

Number of Males	Frequency	Number of Males	Frequency
19	3	28	76
20	4	29	62
21	6	30	59
22	6	31	36
23	18	32	15
24	29	33	11
25	47	34	7
26	49	35	2
27	69	36	1

Table 2

3. In the previous question, what probability did we use in the simulation for the chance that a man would pass? Why does that probability make sense?

How Typical Are Our Households' Ages? The Chi-Square Test

Using chi-square to show that (binned) age distributions are different.

Suppose we had a class picnic and all the people in everybody's household showed up. Would their ages be representative of the ages of all Americans? Probably not. After all, this is not a random sample. But how unrepresentative are the ages? We'll see how to figure that out.

Question

Is the age distribution of the people from the households in your class typical of that of all residents in the United States?

Objectives

In this activity, you will learn to construct a chi-square distribution that can be used to answer questions such as that proposed here.

Prerequisites

You should know how to display a frequency distribution on a dot plot. In addition, although we review this in the following pages, it is helpful if you know how to use a random number table to simulate drawing samples of size 40 from a population divided into three categories of 22%, 28%, and 50%.

Activity

1. Would you expect the age distribution from the households in your class to be typical of the age distribution of all U.S. residents? Why or why not?

2. Fill in Table 1 with data from your class. Your instructor will help you get data for about 40 people—however many households it takes.

Age	Proportion in U.S. Population	Number Observed in Households from This Class	Expected Number in Your Class (national proportion times the observed total)
0–14	0.22		
15–34	0.28		
35+	0.50		
Total	1.00		

Table 1

You would not expect the numbers in the last two columns to match exactly, even if households from your class were drawn randomly from the U.S. population. But are the numbers observed in your class far enough away from those expected to convince you that your households are not representative of the population?

3. The chi-square statistic is a measure of how different the observed column is from the expected column. For your table, compute the chi-square statistic,

$$\chi^2 = \Sigma \frac{(O - E)^2}{E}$$

Is the chi-square statistic that you just computed large enough to convince you that the ages of the households from your class aren't similar to those of a typical random sample of U.S. residents? To answer this question, you will simulate drawing random samples of ages from the population of U.S. residents and see how often you get a value of chi-square that is as large as or larger than the one from your class.

4. Use a random number table to take a random sample of the ages of 40 (or however many you have) U.S. residents. Because 22% of the people are 14 or younger, let the 22 pairs of digits 00, 01, 02, . . . , 21 represent people whose age is 14 or younger.
   a. What pairs of digits will represent people aged 15 to 34?
   b. What pairs of digits will represent people aged 35 and older?

c. Using the first 40 (or so) pairs of digits in the lines of the table assigned to you, take a random sample of the ages of 40 (or so) U.S. residents. Mark your results in Table 2.

Age	Random Digits	Number Observed in Your Sample *(O)*	Number Expected in Your Sample *(E)* (national proportion times the observed total)
0–14	00–21		
15–34			
35+			
Total	1.00		

Table 2

d. Compute the value of chi-square for your sample.

e. Does a large value of chi-square mean that the proportions in the sample are relatively close or relatively far from the proportions in the population?

f. Place an X representing your value of chi-square above a number line that extends from 0 to 15.

g. Repeat this simulation with other members of your class until you have at least 100 values above the number line.

5. What is your estimate of the probability that a random sample of 40 U.S. residents have a chi-square value as large as or larger than that from your class's households? Is the distribution of ages of the households from your class similar to those of a typical random sample of U.S. residents? What do you conclude?

Wrap-Up

1. People sometimes say that "b" and "c" answers occur most frequently on multiple-choice tests. In this activity, you will decide whether there is any evidence of this on the basis of the answer form from the verbal section of a real SAT. (This form was selected randomly from The College Board, 10 SATs, New York: College Entrance Examination Board, 1988.) The correct answers are given in Table 3.

1. d	15. c	29. e	43. a	57. e	71. a
2. d	16. d	30. b	44. a	58. d	72. c
3. b	17. a	31. d	45. b	59. c	73. b
4. b	18. c	32. d	46. e	60. b	74. d
5. c	19. c	33. b	47. d	61. b	75. e
6. e	20. b	34. e	48. b	62. d	76. a
7. b	21. b	35. e	49. d	63. e	77. c
8. a	22. b	36. c	50. b	64. b	78. c
9. a	23. c	37. e	51. a	65. d	79. d
10. b	24. b	38. c	52. a	66. e	80. d
11. c	25. c	39. d	53. c	67. b	81. b
12. b	26. a	40. e	54. c	68. d	82. d
13. e	27. c	41. e	55. a	69. c	83. d
14. e	28. e	42. a	56. c	70. b	84. c
					85. b

Table 3

a. Make a table showing the observed number of answers for each of "a," "b," "c," "d," and "e" and the expected number of answers, assuming they are equally likely to occur. Compute the chi-square statistic for this table.

b. The distribution shown in Figure 1 is the result of computing 1000 values of chi-square for samples taken randomly from a population with equal numbers of answers "a," "b," "c," "d," and "e."

Each dot represents 5 points.

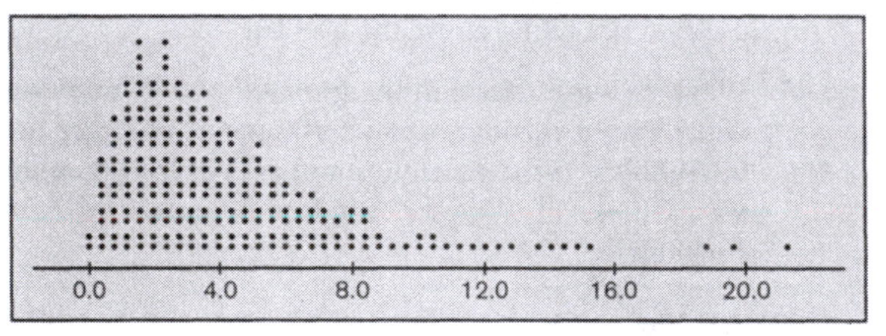

Figure 1

What do you conclude about whether the answers "a," "b," "c," "d," and "e" occur with the same frequency on this type of test? What assumption are you making?

2. Write instructions for students who were absent today, explaining how they could test to see if a die is fair.

Extensions

You can use statistical software to construct the distribution of the chi-square values for a table with varying numbers of rows. When you do this, you may have to give the chi-square function that you use a number of "degrees of freedom," often abbreviated *df.* This is one less than the number of rows.

1. Use statistical software to construct a distribution of chi-square values for tables with various numbers of rows (degrees of freedom). Describe how these distributions change as the number of rows increases.

2. Gregor Mendel performed experiments with peas to test his genetic theory. It has been said that the experimental results were too close to his theory. In one experiment Mendel predicted that he should get 9/16 round yellow peas, 3/16 round green peas, 3/16 wrinkled yellow peas, and 1/16 wrinkled green peas. He writes that he found 315 round yellow peas, 108 round green peas, 101 wrinkled yellow peas, and 32 wrinkled green peas in a sample.
   a. What is the chi-square value for this experiment?
   b. Use software to construct an appropriate distribution of chi-square values for this experiment.
   c. What do you conclude?

Assessment Questions

1. What would be the chi-square value if the proportions in the sample were exactly equal to the proportions in the population? Explain.

2. In 2002, according to the M&M's Web site (http://www.mms.com), the proportions of "plain" milk chocolate M&M's were as follows: 30% browns, 20% each of yellows and reds, and 10% each of oranges, greens, and blues. Your instructor has a bag of plain M&M's. Describe how you would decide, based on the proportions in the bag, whether or not you think the distribution of colors is still the same as it was then. If you have access to statistical software or a powerful calculator, construct the appropriate distribution and state your conclusion.

Coins on Edge

The power of a hypothesis test increases as the sample size increases.

People often flip a coin to make a "random" selection between two options, for example, to choose which team goes first in a competition. This depends on the assumption that the probability of getting heads is 0.5, that is, that the coin is "fair."

Question

Suppose you stand a penny on edge and then make it fall by tapping the table gently. What is the probability that you will get heads?

Objectives

In this activity, you will develop an understanding of the concept of power in a statistical test. Power is the ability of a test to reject a null hypothesis when it is not true.

Prerequisites

You should be familiar with the idea of a hypothesis test. It is possible to do the activity without performing the statistical analysis (of finding the p values and so forth). However, to use the activity as it is written, you need to know what a p value is, how to find binomial probabilities, and how to use the normal approximation to the binomial.

Activity

You need 10 pennies and a flat table. Stand two of the pennies on edge on the table. This may take a while—be patient. (Hint: You might find it easiest to make a penny stand on edge if you place it so that Lincoln is upside-down.) Then, with a gentle downward stroke (not sideways), strike the table so that the coins fall over. Don't pound on the table; just strike it hard enough that the pennies fall over. If they don't fall, hit the table again—with a downward stroke. Count the number of heads facing up; record your results in the first row of Table 1. Then record your feelings about the question "Is the probability of getting heads 0.5?" Write down how sure you feel about your answer to the question.

Now take two more pennies, stand them on edge, and strike the table to make them fall. Record your results in the second row of the table. Also record your feelings about the question "Is the probability of getting heads 0.5?" Do you feel that you are now sure of the answer to the question?

Repeat this experiment for each row of the table, using the number of pennies indicated in the first column. Be sure to record your reaction after each repetition. When you have completed the activity, you will have a total of 50 individual results. After each repetition, record not only the number of heads and the cumulative number of heads but also what conclusions you feel you can draw regarding the question "Is the probability of getting heads 0.5?" and how sure you feel about your answer.

Number of Pennies	Number of Heads	Cumulative Heads	Cumulative Pennies	Thoughts and Remarks to This Point
2			2	
2			4	
2			6	
2			8	
2			10	
5			15	
5			20	
5			25	
5			30	
10			40	
10			50	

Table 1: Data recording sheet

Statistical Analysis

Use the data from the first row of the data sheet to perform a test of the null hypothesis that the probability of getting heads is 0.5; use a general (two-sided) alternative hypothesis. You have only $n = 2$ observations, so you will want to use the binomial distribution to find the p value. Note that the p value is the probability of getting data at least as extreme as the data you obtained, under the assumption that probability of getting heads is indeed 0.5. That is, the p value is P(sample percentage heads would differ from 0.5 by at least as wide a margin as obtained in this sample, if we were to repeat the experiment). If $\alpha = 0.05$, do you reject H_0?

Repeat the hypothesis test, but now use the data from the second row of the data sheet in Table 1. That is, do a test based on $n = 4$ observations.

Repeat the hypothesis test for each row of the data sheet. When n is large (say, starting with $n = 25$), you might want to use the normal approximation to the binomial distribution when finding the p value. At what point do you reject H_0? That is, how large a sample do you need before you can reject H_0?

Wrap-Up

1. Summarize how the p value is related to the sample size. Give an explanation of this phenomenon.
2. Write a brief summary of what you learned in this activity about how power—the ability to reject H_0—is related to sample size.

Extensions

Consider the data for 50 trials from just one student or group. As a class, you could construct a confidence interval for the true probability, p, of getting heads. One way to approach this is to ask yourselves, "Could p be 0.5?" This question will already have been answered (as "no") at the end of the activity. Then ask, "Could p be 0.6?" You can do a quick z-test of this hypothesis by finding the z-score and using the rule "reject H_0 if the z-score is outside the range ± 2."

Now consider other possible values of p. If you have nine groups of students, for example, then ask one student or group to check whether p might be 0.55, another group to check whether p might be 0.60, and so on up to 0.95. The values for which $-2 < z\text{-score} < 2$ (that is, the answers to the hypothesis-testing question is "yes, this might be the true value of p") are inside the confidence interval and other values are outside the confidence interval.

You can refine the confidence interval by checking values of p that are "in the cracks" at the edge of the interval. For example, using the data in Table 1 (40 out of 50 heads), we get $|z\text{-score}| < 2$ for $p = 0.70, 0.75, 0.80$, and 0.85. Checking values of p between 0.65 and 0.70, we get $|z\text{-score}| < 2$ for $p = 0.67, 0.68$, and 0.69. Checking values of p between 0.85 and 0.90, we get $|z\text{-score}| < 2$ for $p = 0.86, 0.87$, and 0.88. Thus, the refined confidence interval for p is (0.67, 0.88).

Assessment Question

Suppose a researcher wants to take a sample of 15 voters to test, using $\alpha = 0.05$, the hypothesis that half of the population of voters approve of the job performance of a certain politician. Suppose that, in fact, 70% of the population approves of this person's job performance. Discuss the researcher's plan and what the researcher might conclude.

V

Projects

Many instructors of introductory statistics have found that student projects enhance motivation, interest, and understanding in the course. Projects require planning and executing a series of steps, over a period of time—even exceeding a month—to solve a particular problem or answer a specific question. Among the advantages of projects are the following:

- Students work on a complete problem, from formulating the question, through gathering and analyzing data, to reporting conclusions.
- Students bring a variety of statistical techniques to bear on one problem and thus develop connections among the topics.
- Students gain experience in using statistics in a realistic way.
- Students often work with partners or in teams, thereby gaining experience in group work.
- Students gain experience in communicating statistical ideas.

It is hard to manage completely open-ended projects in a one-semester or one-quarter introductory course. Neither the instructor nor the student may ever see the finished product. Thus, it helps to focus projects a little, concentrating on the main themes in introductory statistics. These themes include data exploration, quality improvement, sample surveys, experiments, and modeling. The first two fit together quite nicely because much of quality improvement work in industry involves careful exploration of data. Thus, we suggest four main themes for projects: exploration of data and improvement in quality, sample surveys, experiments, and modeling. The following sections provide some details on the statistical ideas embedded in these themes, with real examples of these ideas in practice. The goal is to provide sound background information that students can draw upon when carrying out a project.

THEME

Exploration of Data and Improvement in Quality

Data and Decisions

In today's information society, decisions are made on the basis of data. A student checks the calorie chart before selecting a fast-food lunch. A homeowner checks the efficiency rating before purchasing a new refrigerator. A physician checks the outcomes of recent clinical studies before prescribing a medication. An engineer tests the tensile strength of wire before winding it into a cable. The decisions made from data—correctly or incorrectly—affect each of us every day. How to systematically study data for the purpose of making decisions is the overarching theme of modern statistics, and the basic ideas on how to make sense out of data are the subject of the first set of activities. You should study these ideas thoroughly in an introductory statistics course because they form the building blocks for the remainder of statistics.

Quality Really Is Job One

"Quality is job one." This slogan is now synonymous with an American automobile manufacturer—but it is far more than a slogan. A commitment to quality helped this manufacturer recapture a sizable share of the automobile market once lost to foreign competition. Similar stories can be told about many other firms whose products have risen in customer satisfaction since the firm began emphasizing quality within the production process (continuous process improvement). Producing a product or service of high quality is a complex matter, but virtually all of the success stories have one thing in common: Decisions were and are made on the basis of objective, quantitative information—data!

To see how the Ford Motor Company emphasized quality, data, and statistics, consider the following statement from its manual on continuous process control.

To prosper in today's economic climate, we—Ford, our suppliers and our dealer organizations—must be dedicated to never-ending improvement in quality and productivity. We must constantly seek more efficient ways to produce products and services that consistently meet customer's needs. To accomplish this, everyone in our organization must be committed to improvement and use effective methods. . . . [T]he basic concept of using statistical signals to improve performance can be applied to any area where work is done, the output exhibits variation, and there is desire for improvement. Examples range from component dimensions to bookkeeping error rates, performance characteristics of a computer information system, or transit times for incoming materials.

Ford is not alone in its emphasis on the effective use of statistics. The chairman/CEO of the Aluminum Company of America has stated

As world competition intensifies, understanding and applying statistical concepts and tools is becoming a requirement for all employees. Those individuals who get these skills in school will have a real advantage when they apply for their first job.

Arno Penzias, vice president for Research at AT&T Bell Laboratories, has written that

The competitive position of industry in the United States demands that we greatly increase the knowledge of statistics among our engineering graduates. Too many of today's manufacturers still rely on antiquated "quality control" methods, but economic survival in today's world of complex technology cannot be ensured without access to modern productivity tools, notably applications of statistical methods (Science, 1989, 244:1025).

Even in our daily lives, all of us are concerned about quality. We want to purchase high-quality goods and services, from automobiles to television sets, from medical treatment to the sound at the local movie theater. Not only do we want maximum value for our dollar when we purchase goods and services, but we also want to live high-quality lives. Consequently, we think seriously about the food we eat, the amount of exercise we get, and the stress that accumulates from our daily activities. How do we make the many decisions that confront us in this effort? We compare prices and value, we read food labels to determine calories and cholesterol, we ask our physician about possible side effects to a prescribed medication, and we make mental notes about our weight gains or losses from week to week. In short, we, too, are making daily decisions on the basis of objective, quantitative information—data!

Formally or informally, then, data are the basis for many of the decisions made in our world. Using data lets us make decisions on the basis of factual information rather than on subjective judgment: Facts should produce better results than fantasy. But whether or not the data lead to good results depends on how we get the data and how we analyze it. For many of the simpler problems we face, the data are readily available. When we evaluate the nutrition in food or decide on the most efficient appliance, we read the label. Thus, techniques for analyzing data will be presented first, with ideas on how to produce good data following later.

A Model for Problem Solving

Whether we are discussing products, services, or lives, improving quality is the goal. Along the path to improved quality, we must make numerous decisions, and we must make them on the basis of objective data. It is possible, however, that these decisions could affect each other, so it is wise to view the set of decisions together in a wider problem-solving context. Consider, for example, a civil engineer who is to solve the problem of hampered traffic flow through a small town with only one main street. After collecting data on the volume of traffic, the immediate decision seems easy—widen the main street from two to four lanes. However, careful thought might suggest that more traffic lights will be needed for the wider street so that traffic can cross it. In addition, traffic now using the side streets will begin using the main street once it is improved. A "simple" problem has become more complex once all the possible factors are brought into the picture. The whole process of traffic flow can be improved only by taking a more detailed and careful approach to the problem.

The central idea is to use data analysis to improve the quality of a process, whether the "process" is moving cars through a town, growing tomatoes, buying a VCR, or studying for an examination. You must examine all aspects of the process, because the goal is to improve it throughout. This may involve the study of many variables and how they interrelate, so a systematic approach to problem solving will be essential to your ability to make good decisions efficiently. In recent years, numerous models for solving problems have evolved within business and industry; the model outlined next contains the essential steps present in all of them.

1. State the problem or question.

   This sounds like an obvious and, perhaps, easy step. But going from a loose idea or two about a problem to a clear statement of the real problem requires careful investigation of the current situation so that one can develop well-defined goals for the study. ("I am pressed to get my homework assignments done on time, and I do not seem to have adequate time to complete all my reading assignments. I still want to work out each day and to spend some time with friends. Upon review of my study habits, it seems that I study rather haphazardly and tend to procrastinate. The real problem, then, is not to find more hours for study but to develop a study plan that is efficient.")

2. Collect and analyze data.

   Now list all factors you think affect the problem and collect data on all factors you think are important. Direct a plan toward solving the specific problem defined in the first step. Use appropriate data analysis techniques, according to the data you collected. ("The data show that I study 5 hours a day and work late into the evening. It also shows that the gym is crowded when I arrive, and this may slow down my workout.")

3. Interpret the data and make decisions.
   After you analyze the data and study your analysis carefully, pose potential solutions to the original problem or question. ("I see that studying late into the evening is often necessary because I watch television and visit with friends before studying. Consequently, I am tired when I begin studying. I will set a schedule that puts my study time early in the evening and my visiting later in the evening. Also, I will work out in the mornings rather than the afternoons to save time.")

4. Implement and verify the decisions.
   Once you pose a solution, put it into practice on a trial basis (if feasible). Collect new data on the revised process to see if improvement is actually realized. ("I tried the earlier study time for 2 weeks, and it worked fine. I seemed to have more time to complete my assignments, even though the data show that I was not devoting any more hours to study. The gym is just as busy in the morning, so I realized no saving of time with the new strategy.")

5. Plan next actions.
   The trial period of step 4 may show that the earlier decision solved the problem. More likely, though, the decision was only partially satisfactory. In any case, there is always another problem to tackle; plan this now, while the whole process is still firmly in mind and the data are still fresh. ("I would still like to find more time for pleasure reading. I will see how I could fit that into my revised schedule.")

Improved Payment Processing in a Utilities Firm

Office personnel responsible for processing customer payments to a large utility company noticed that they were receiving complaints both from customers ("Why hasn't my check cleared?") and from the firm's accounting division ("Why did it take so long to get these receipts deposited?"). The office staff decided it was time to swing their quality improvement training into action. The following is a brief summary of their quality improvement story.

1. State the problem or question.
   After brainstorming on the general problem, the team collected background data on the elapsed time for processing payments and found that about 2% of the payments (representing 51,000 customers) took more than 72 hours to process. They suspected that many of these were "multis" (payments containing more than one check or bill) or "verifys" (payments in which the bill and the payment do not agree). Figure 1 is a *Pareto chart* that demonstrates the correctness of their intuition; multis account for 63% of the batches requiring more than 72 hours to process, but they are only 3.6% of the total payments processed. Verifys are not nearly as serious as first suspected, but they are still the second leading contributor to the problem. The problem can now be made specific; concentrating first on the multis, reduce payments requiring more than 72 hours of processing time to 1% of the total.

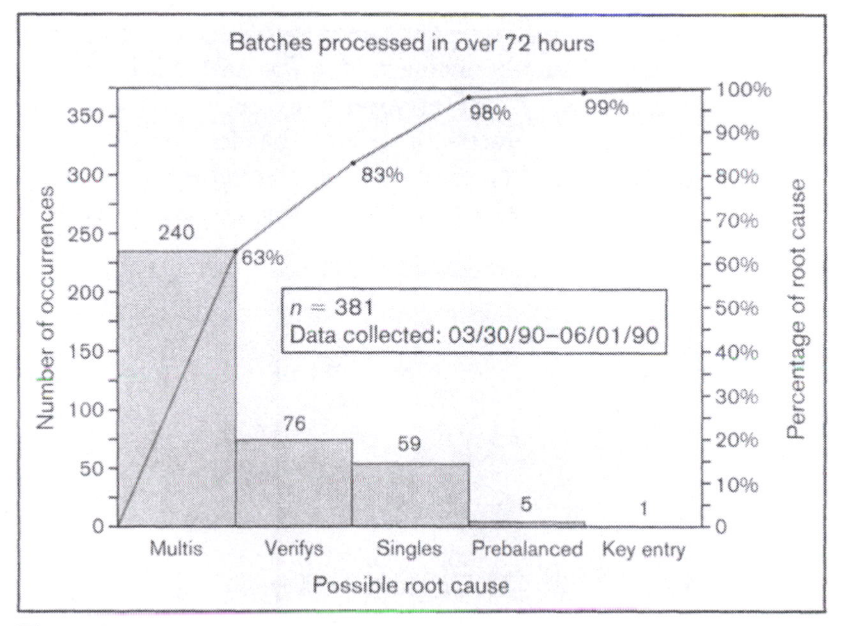

Figure 1

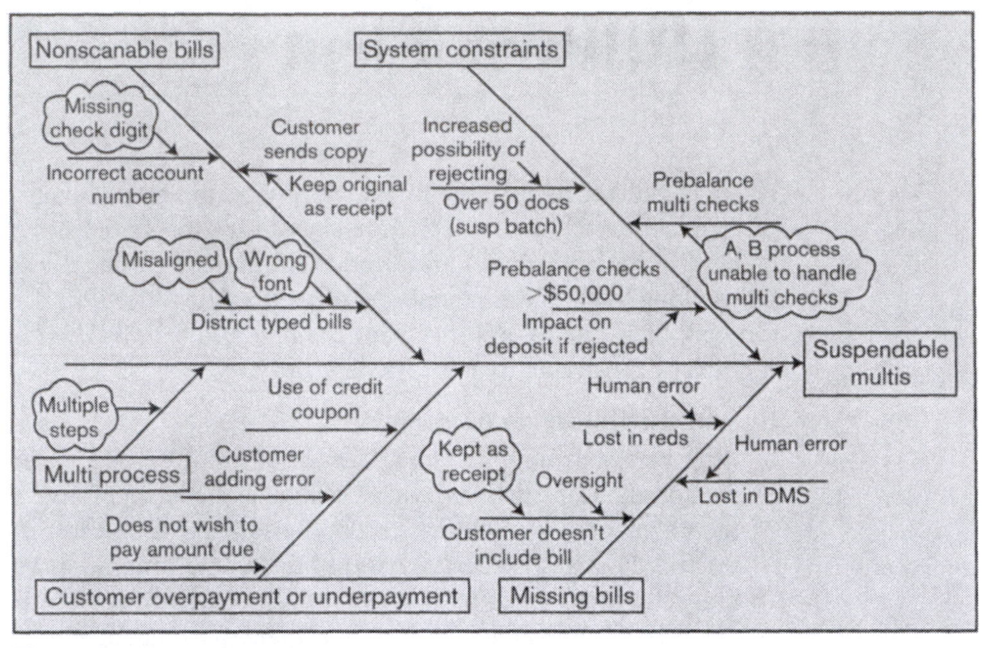

Figure 2

2. Collect and analyze data.

   What factors affect the processing of multis? A cause-and-effect diagram, shown in Figure 2, lists all important factors and demonstrates how they might relate to each other. A missing check digit on a bill leads to an incorrect account number, which, in turn, causes a nonscanable bill. Sometimes a customer keeps the bill as a receipt and, as a result, the check cannot be processed. The largest single factor that was possible to correct, in this case, was the fact that the computerized auto-balancing process was unable to handle multiple checks.

   Another factor, cash carryover from the previous day, is shown to affect the entire payment-processing system. The effect on the elapsed time of processing payments can be seen in the scatter plot of Figure 3. An efficient system for handling each day's bills must reduce the carryover to the next day.

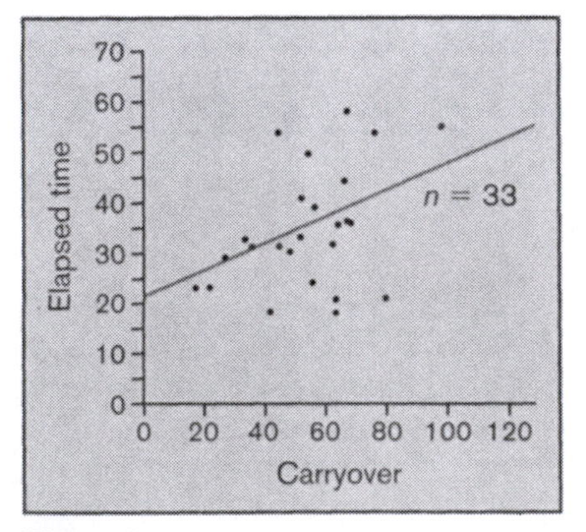

Figure 3

3. Interpret the data and make decisions.
   With data in hand that clearly show the computerized check-processing system to be a major factor in the processing delays, the staff developed a plan to reprogram the machine so that it would accept multiple checks and coupons as well as both multis and verifys that were slightly out of balance.

4. Implement and verify the decisions.
   The reprogramming was completed, and along with a few additional changes, it produced results that were close to the goal given in the problem statement. The record of processing times during a trial period are shown in the time series plot of Figure 4.

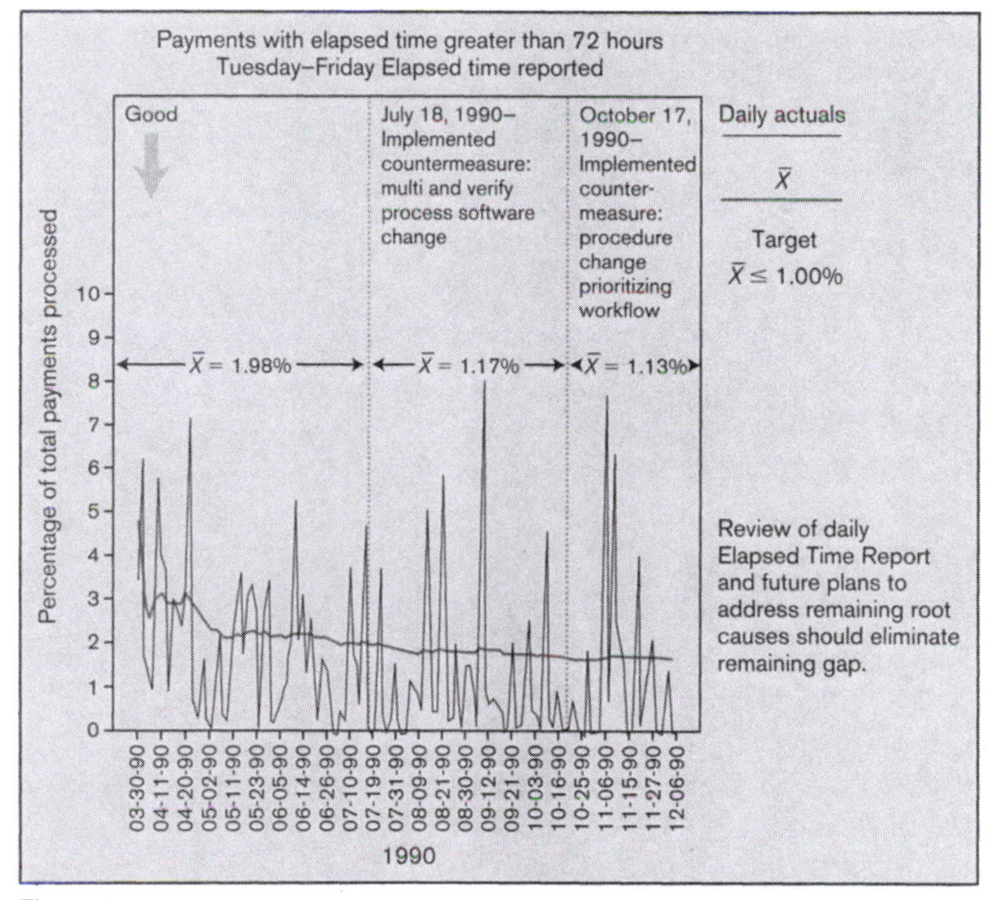

Figure 4

5. Plan next actions.
   The carryover problem remains, and along the way, it was discovered that the mail-opening machine has an excessive amount of downtime. Resolution of these two problems led to an improvement that exceeded the goal set in the problem statement.

On to Real Data Collection and Analysis

The heart of statistics is collecting and analyzing data. Data do not simply appear; the correct analysis is not always obvious. These two arms (collection and analysis) must work together toward the common goal of solving a problem. To achieve this goal, you must collect data according to a plan. The plan used within the quality improvement programs now meeting with great success in industry is a good one to follow in almost any setting that requires problem solving. The theme of improving the quality of a process gives you a general framework for exploring data; it gives order and direction to statistical applications at any level. Projects in *Exploration of Data and Improvement in Quality* give students practical experience in using the basic statistical tools generally found essential to solving real quantitative problems.

Sample Survey

What Is a Sample Survey?

Your college administrators want to know how many students will want parking spaces next year. How can we get reliable information on this question? One way is to ask all of the returning students, but even this procedure would be somewhat inaccurate (why?) and very time-consuming. We could take the number of spaces in use this year and assume next year's needs will be about the same, but this will result in inaccuracies as well. A simple technique that works very well in many cases is to select a sample from those students who will be attending the school next year and ask each of them if they will be requesting a parking space. From the proportion of "yes" answers, we estimate the number of spaces we will need.

The scenario just outlined has all of the elements of a typical sample survey problem. There is a question of "how many?" or "how much?" to be determined for a specific group of objects called a *target population,* and an approximate answer is to be derived from a sample of data extracted from the population of interest. The approximate answer will be a good approximation only if the sample truly represents the population under study. Randomization plays a vital role in selecting samples that truly represent a population and, hence, produce good approximations. It is clear that we would not want to sample only our classmates or friends on the parking issue. It is less clear, but still true, that virtually any sampling scheme that depends on subjective judgments as to who should be included will suffer from sampling bias.

Once we know who is to be in the sample, we still need to get the pertinent information from them. The method of measurement, that is, the questions or measuring devices we use to obtain the data, should be designed to produce the most accurate data possible and should be free of measurement bias. We could choose a number of ways to ask students about their parking needs for next year. Here are a few suggested questions:

- Do you plan to drive to school next year?
- Do you plan to drive to school on more than half of the school days next year?
- Do you have regular access to a car for travel to school next year?
- Will you drive to school next year if the cost of parking increases?

Which of these questions do you think might bias the results, and in which direction? A few moments of reflection should convince you that measurement bias could be serious in even the simplest of surveys.

Because it is so difficult to get good information in a survey, every survey should be pretested on a small group of subjects similar to those that could arise in the final sample. The pretest, or pilot test, not only helps improve the questionnaire or measurement procedures but also helps determine a good plan for data collection and data management. For example, can we list a mutually exclusive and exhaustive set of meaningful options on the parking question so that we can easily code responses for the data analysis phase? The data analysis should lead to clearly stated conclusions that relate to the original purpose of the study. The goal of the parking study is to focus on the number of spaces needed, not the types of cars students drive or the fact that auto theft may be a problem.

We will discuss and illustrate key elements of any sample survey in this section. These elements include the following:

1. State the *objectives* clearly.
2. Define the *target population* carefully.
3. Design the *sample selection* plan using *randomization* to reduce *sampling bias.*
4. Decide on a method of measurement that will minimize *measurement bias.*
5. Use a *pretest* to try out the plan.
6. Organize the data collection and *data management.*
7. Plan for careful and thorough *data analysis.*
8. Write *conclusions* in the light of the original objectives.

We will now turn to some examples of sample surveys. You can find results from surveys like these in the media on almost any day.

A Typical Election Poll

The article reprinted here from the *Gainesville Sun* appeared close to the end of the presidential race among Bush, Clinton, and Perot in 1992. Read the article carefully before studying the material presented in the following paragraphs. As you go through this discussion, make sure all of the statements are justified and try to answer the open questions left for the reader.

A Typical Election Poll

Undecided will sway Fla. vote

BY BILL RUFTY
NYT Regional Newspapers

LAKELAND—The fate of Florida's 25 electoral votes apparently lies in the hands of the 3 percent of the voters who still don't know who they will vote for, according to results from the latest Florida Opinion Poll.

Those few voters will break the deadlock between Republican President George Bush and Democratic challenger Bill Clinton in Tuesday's presidential election.

While independent candidate Ross Perot does not have a chance to capture the state, he apparently has a tight grip on about a fifth of the vote. The poll showed that:

- 39 percent support or are leaning toward Bush.
- 37 percent support or are leaning toward Clinton.
- 21 percent support or are leaning toward Perot.
- 3 percent are undecided.

The Florida Opinion Poll, which is sponsored by The New York Times newspapers in Florida, contacted state residents in a random telephone sample.

Along the political spectrum, Bush is heavily dependent on conservatives in Florida, while Clinton captures nearly half the moderate voters and nearly three-quarters of the liberals.

Presidential race in Florida too close to call

If the presidential election were today would you vote for George Bush, Bill Clinton or Ross Perot?

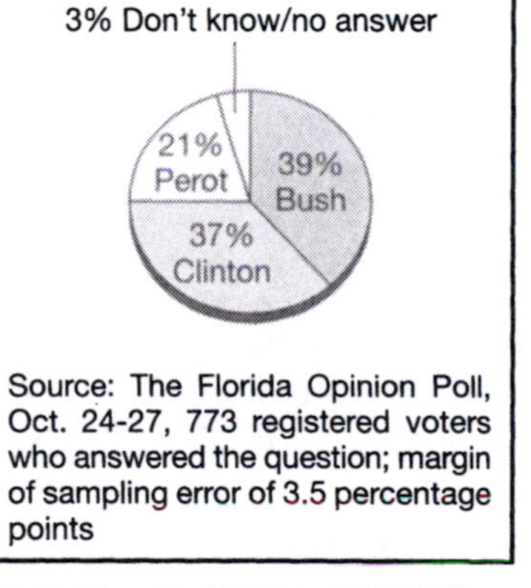

Source: The Florida Opinion Poll, Oct. 24-27, 773 registered voters who answered the question; margin of sampling error of 3.5 percentage points

Adapted from Mark Williams/NYTRENG Graphics Network

Perot pulls almost identical support from liberals and conservatives, while drawing a quarter of moderates.

The poll found that among conservative voters:

- 63 percent back Bush.
- 18 percent back Clinton.
- 19 percent back Perot.

Among moderate voters:

- 28 percent back Bush.
- 46 percent back Clinton.
- 26 percent back Perot.

Among liberal voters:

- 11 percent back Bush.
- 71 percent back Clinton.
- 18 percent back Perot.

While Perot pulls support from across the political spectrum, he is taking away more of Bush's Republicans than Clinton's Democrats.

Among Republicans:

- 64 percent back Bush.
- 12 percent back Clinton.
- 22 percent back Perot.
- 2 percent were undecided.

But Perot takes Democratic votes from Clinton, but not as many, and there are more registered Democrats in Florida than Republicans.

(continued)

Among Democrats:

- 20 percent back Bush.
- 61 percent back Clinton.
- 17 percent back Perot.
- 2 percent were undecided.

Of the 6.5 million registered voters in Florida, 3.3 million, or 51 percent, are registered as Democrats, while 2.7 million voters, or 41 percent, are registered as Republicans and 550,292, or 8 percent, are registered as independents or members of third parties.

By age, candidates seem to appeal equally across all age groups, with each drawing in the 30 percent range, except for the 45- to 64-year-olds.

By gender, there appears to be no major difference in the number of women or men going for one candidate more than the others.

Source: Gainesville Sun, October 30, 1992.

How Polls Were Conducted

NYT Regional Newspapers

The latest Florida Opinion Poll was conducted by telephone from Oct. 24 to 27 with 773 voters considered most likely to go to the polls on Election Day.

The telephone numbers used in the survey were formed at random by a computer programmed to ensure that each area of the state was represented in proportion to its population.

The results based on responses from all 773 most likely voters have a margin of sampling error of 3.5 percentage points. That means if the New York Times newspapers in Florida asked every voter in the state the same questions, in most cases, the results would be within 3.5 percentage points of the results obtained by the survey.

Interviewers used a series of three questions to determine voters who were most likely to go to cast their ballots Tuesday.

In questions where only the answers of smaller groups are used, the margin of sampling error is larger.

For example, the margin of sampling error for just registered Democrats or only registered Republicans will be higher.

In addition to sampling error, the practical difficulties of conducting any poll can induce other forms of error.

The poll attempts to answer the question of who would win the presidential race if the voting took place on October 30, 1992. The specific question asked was, "If the presidential election were today, would you vote for George Bush, Bill Clinton, or Ross Perot?" Secondary questions dealt with subgroups of voters, such as conservatives versus liberals and Republicans versus Democrats. This study is clearly a sample survey because its goal is simply to describe the proportion of voters who plan to vote for the various candidates. (Essentially, the poll is seeking to find out "how many" people will vote for each candidate.)

The target population under study here is not the entire population of Florida or even the entire population of registered voters in Florida. It is the *population of voters most likely to vote on election day.* Whether a voter fell into this population was determined by a series of questions, which are not presented in the article. (Speculate as to what you think these questions might have been.) The sample was selected by randomly selecting telephone numbers from different areas of the state. (How do you think this randomization might have been accomplished? Why was the state divided into areas, with randomization done within the areas?) The sample size was 773. (How do you think they arrived at such a strange number?)

The measurements (simple statements of opinion) for this study were collected through telephone conversations. The question asked was quite straightforward, so measurement bias does not appear to be a problem. The data are nicely summarized in proportions and presented on a pie chart with useful information, such as the sample size and margin of sampling error, included in the box containing the chart. Little analysis is presented beyond this summary. The principal conclusion is that the race between Bush and Clinton is too close to call. Why would the pollsters make that conclusion when it is clear that Bush has the highest percentage of votes in the sample? To understand this, we need to delve a little more deeply into the margin of sampling error. Sample proportions will vary, from sample to sample, according to a definite and predictable pattern (in the long run). In fact, 95% of all possible sample proportions, for samples of size n, that could be obtained from a single population will fall within 2 standard deviations of the true population proportion. If a sample proportion is denoted by p, then the standard deviation of the possible sample proportions is given by

$$\sqrt{\frac{p(1-p)}{n}}$$

and 95% of the potential sample proportions will fall within

$$2\sqrt{\frac{p(1-p)}{n}}$$

of the true population proportion. This 2-standard-deviation interval is called the *margin of sampling error.* Substituting 0.39 for p and 773 for n results in a calculated margin of error equal to 0.035. Thus, the true proportion of voters favoring Bush may well be anywhere in the interval $0.39 - 0.035$ to $0.39 + 0.035$, or (0.355, 0.425). (Find the corresponding interval for Clinton. Do you see why the pollsters did not want to call this race? Find the corresponding interval for Perot. Why did the writers of this article report only one value for the margin of error?)

Notice that the sample size is in the denominator of the margin of sampling error formula. Thus, the pollster could have reduced the margin of error by increasing the sample size (why?). Now, look back at one of the surveys you have conducted earlier. Find the margin of sampling error for a proportion and then find the sample size that would be required to cut this margin of error in half.

Let's look back at the article carefully to see what else we might like to know. Much discussion is presented on the breakdowns of the percentages among conservative, moderate, and liberal voters. Is there any way to attach a margin of sampling error to these percentages? If we could find these margins of error, would they be larger or smaller than 0.035? How about the breakdown between Republicans and Democrats? Can we attach a meaningful margin of error to these percentages?

It would be nice to know more about the sampling scheme. How was the randomization of telephone numbers conducted? What areas of the state were used? Why a sample size of 773? (This could indicate that some who were called refused to answer the poll. Could this cause any bias?) What were the questions that determined whether a response was used? (Could these questions cause any bias?)

You might want to look at the Florida vote to see if the pollsters were right in implying that the race would be close. Also, you might want to study polls of current interest in a similar fashion.

We should point out that the 2000 presidential results in Florida—in the race between George W. Bush, Al Gore, Ralph Nader, and others—was spectacularly noteworthy. Yet the problem and controversy had nothing to do with sampling error. It is interesting to ask whether, and to what extent, experience with and knowledge of statistics and sample surveys help us understand what happened there.

The Nielsens

Almost everyone watches television and, as a result, has some awareness that what is available to watch is determined by the Nielsen ratings. A show that does poorly in the Nielsens is not going to be on a major network very long (why not?). The following article shows the final Nielsen ratings for a week in March 1995. Read the explanation in this article before reading the following discussion.

NBC Sitcoms Still Dominate Thursday Night

The Nielsens

SCOTT WILLIAMS
The Associated Press

NEW YORK—The Peacock can strut again.

For the fifth consecutive week, NBC won the prime time ratings crown behind top-rated "Seinfeld" and Top 10 performances from four other shows in its Thursday lineup.

For the week, NBC averaged an 11.5 rating and a 19 percent audience share.

ABC, the season-to-date front-runner, finished second with an 11.1 rating, 19 share. CBS was third, with a 9.2 rating, 16 share.

Top 20 listings include the week's ranking, with rating for the week, season-to-date rankings in parentheses, and total homes.

An "X" in parentheses denotes one-time-only presentation. A rating measures the percentage of the nation's 95.4 million TV homes. Each ratings point represents 954,000 households, as estimated by Nielsen Media Research.

1. (1) "Seinfeld," NBC, 21.4, 20.4 million homes
2. (2) "Home Improvement," ABC, 20.5, 19.6 million homes
3. (3) "E.R.," NBC, 19.8, 18.9 million homes
3. (11) "Friends," NBC, 19.8, 18.9 million homes
5. (4) "Grace Under Fire," ABC, 19.6, 18.7 million homes
6. (7) "NYPD Blue," ABC, 16.9, 16.1 million homes
6. (5) "60 minutes," CBS, 16.9, 16.1 million homes
8. (12) "Mad About You," NBC, 15.3, 14.6 million homes
9. (10) "Hope & Gloria," NBC, 14.9, 14.2 million homes
10. (16) "Murphy Brown," CBS, 14.8, 14.1 million homes
11. (9) "Murder She Wrote," CBS, 14.3, 13.6 million homes
11. (17) "20-20," ABC, 14.3, 13.6 million homes
13. "McBain's 87th Precinct—NBC Sunday Movie," 13.8, 13.2 million homes
14. "Chicago Hope," CBS, 13.7, 13.1 million homes
15. (14) "Ellen," ABC, 13.5, 12.9 million homes
16. (20) "Dave's World," CBS, 13.4, 12.8 million homes
16. (17) "Awake To Danger—NBC Monday Movies," 13.4, 12.8 million homes
18. (8) "Roseanne," ABC, 13.3, 12.7 million homes
19. (22) "Betrayed–ABC Sunday Movie," 13.2, 12.6 million homes
20. "Cybill," CBS, 13.1, 12.5 million homes
21. "Thunder Alley," ABC, 12.7
22. "Step by Step," ABC, 12.5
23. "Primetime Live," ABC, 12.4
24. "Frasier," NBC, 12.3
24. "Full House," ABC, 12.3
26. "Family Matters," ABC, 12.1
27. "Boy Meets World," ABC, 12.0
28. "America's Funniest Home Video," ABC, 11.8
29. "Law and Order," NBC, 11.7
29. "Nanny," CBS, 11.7
31. "Dateline NBC," (Tuesday) 11.5
31. "Dateline NBC," (Wednesday) 11.5
33. "On Our Own," ABC, 11.3
34. "Wings," NBC, 11.1

(continued)

34. "Beverly Hills, 90210," Fox, 11.1
36. "Return-TV Censored Bloopers," NBC, 10.9
37. "Lois & Clark," ABC, 10.8
38. "John Larroquette Show," NBC, 10.9
39. "Far and Away—ABC Monday Movie," 10.4
40. "Something Wilder," NBC, 10.2
41. "Melrose Place," Fox, 10.0
42. "Dateline NBC," (Friday) 9.9
43. "Fresh Prince of Bel Air," NBC, 9.7
44. "Coach," ABC, 9.6
44. "Unsolved Mysteries," NBC, 9.6
46. "Walker, Texas Ranger," CBS, 9.5
47. "America's Funniest Home Videos," ABC, 9.4
48. "All American Girl," ABC, 9.3
48. "Cosby Mysteries," NBC, 9.3
48. "Peter Jennings Reporting," ABC, 9.3
51. "Under One Roof," CBS, 9.2
52. "Simpsons," Fox, 9.1
53. "Blossom," NBC, 9.0
53. "Sister, Sister," ABC, 9.0
55. "Greatest Commercials," CBS, 8.8
56. "George Wendt Show," CBS, 8.7
56. "Living Single," Fox, 8.7
58. "Northern Exposure," CBS, 8.4
58. (X) "Mommies," NBC, 8.4
60. "Empty Nest," NBC, 8.3
60. "Married . . . With Children," Fox, 8.3
60. "X-Files," Fox, 8.3
63. "The Dead Pool—CBS Tuesday Movie," 8.2
64. "America's Most Wanted," Fox, 8.1
64. "Seaquest DSV," NBC, 8.1
66. "Extreme," ABC, 8.0
66. "Mommies," NBC, 8.0

Source: Gainesville Sun 3/26/95

Of the 95.4 million households in the United States, Nielsen Media Research randomly samples 4000 on which to base their ratings. This is accomplished by randomly selecting city blocks (or equivalent units in rural areas), having an enumerator actually visit the sampled blocks to list the housing units, and then randomly selecting one housing unit per block. These sampled housing units are the basic units for all of the ratings data. After a housing unit is selected, an electronic device is attached to each TV set in the household. This device records when the set is turned on and the network to which it is tuned. Information from the network determines which show is actually playing at any point in time. This device gives information on what is happening to the TV set, but it doesn't tell who or how many people are viewing the programs. For this information, Nielsen must rely on individuals in the household recording when they personally "tune in" and "tune out."

The rating for a program is the percentage of the sampled households that have TV sets on and tuned to the program in question. (Note that in the estimated ratings, the denominator of the sample proportion is always 4000.) So, a rating is an estimate of the percentage of households viewing a particular program. A share for a program is an estimate of the percentage of viewing households that have a TV tuned to that particular program, where a viewing household is one for which at least one TV set is turned on. (Will the denominator of the estimated shares be 4000, greater than 4000, or less than 4000?)

In reality, the ratings and shares are slightly more complicated than explained here. A rating for any program is taken minute by minute and then averaged over the length of the program. This attempts to adjust for the fact that not all viewers watch all of a program. Thus, the final rating for *Seinfeld* would be the average of all ratings taken over the half-hour duration of the show, while the final rating for a basketball game would be the average of ratings taken over the entire time (perhaps a couple of hours) that the game was on the air.

Now, look over the article once again. Discuss any points that are misleading or unclear. Specifically, discuss the following:

1. Why are the shares always greater than the ratings?
2. Are there sources of potential bias in the data collection plan?
3. Can a margin of sampling error be approximated for a rating?
4. Can a margin of sampling error be approximated for a share?
5. Read the following article on computer vision research and relate this to the points made earlier. How could a computer vision device improve the Nielsens?

UF Computer Vision Researchers Develop "Smart" Technology For Nielsen's People-Watching TV

The next time you sit down to watch television, don't be surprised to find that your television is watching you.

At least that's what A.C. Nielsen Media Research is planning.

Transferring cutting-edge "smart" technology developed at UF's Computer Vision research Center, Nielsen hopes to employ a passive "peoplemeter" to silently and automatically record who is watching what, when and where in 4,000 Nielsen homes across the country.

"This passive peoplemeter means that, within the next two years, the demographic information concerning television viewing that we pass on to our clients will be much more accurate," said Jo LaVerde, Nielsen's director of communications.

The device Nielsen now uses requires volunteers to identify themselves with the push of a button. But many viewers, especially children, forget to log on. As a result, researchers claim that only about 50 percent of data collected now is usable.

The peoplemeter uses computer image recognition, an emerging technology that uses lasers to follow moving images and a computer that recognizes those images—whether they are people, pets or enemy tanks.

"The applications for this kind of technology are limitless, for both the military and civilian sectors," said UF computer and information sciences Professor Gerhard Ritter, who is heading the research at the Computer Vision Research Center. "We're talking about very Tom Clancy stuff here."

For Nielsen, more accurate demographic information means greater profits for their clients, including those in the advertising industry, which alone spends more than $30 billion a year on television advertising.

But computer recognition systems, like the human eye, can be easily fooled by camouflage or disguise.

"If it's easy to fool a human, just imagine how easy it is to fool a machine," said Ritter. "We've only just discovered the tip of the iceberg with this technology."

Ritter hopes to eventually create an image recognition system that acts like the human eye and responds to visual stimuli the same way the brain does, making it as foolproof as possible.

Ritter's computer vision research is supported by more than $3 million in National Science Foundation and Department of Defense grants. Another $100,000 comes from A.C. Nielsen. Ritter expects to finish a prototype passive peoplemeter in two years.

Source: Alligator, 11/25/92

The University of Florida is an Equal Opportunity/Affirmative Action Institution

Experiment

What Is an Experiment?

How many times has a parent or a teacher admonished a student to turn off the radio while doing homework? Does listening to music while doing homework help or hinder? To answer specific questions like this, we must conduct carefully planned experiments. Let's suppose we have a history lesson to study for tomorrow. We could have some students study with the radio on and some study with the radio off. But the time of day that the studying takes place could affect the outcomes as well. So, we have some study in the afternoon with the radio on and some study in the afternoon with the radio off. Other students study in the evening, some with the radio on and some with it off. The measurements on which the issue will be decided (for now) are the scores on tomorrow's quiz.

Because males might produce different results from females, perhaps we should control for sex by making sure that both males and females are selected for each of the four treatment slots. On thinking about this design for a moment, we conclude that the native ability of the students might have some affect on the outcome as well. All of the students are from an honors history course, so it's difficult to differentiate on ability. Therefore, we will randomly assign students (of similar ability) to the four treatment groups in the hope that any undetected differences in ability will balance out in the long run.

The preceding outline of a study has most of the key elements of a designed experiment. The goal of an experiment is to measure the effect of one or more treatments on experimental units appropriate to the question at hand. Here, there are two main treatments: the radio and the time of day that study occurs. Another variable of interest is the sex of the student (the experimental unit in this case), but this variable is directly controlled in the design by making sure we have data from both sexes for all treatments. The variability "ability" cannot be controlled as easily, so we randomize the assignment of students to treatments to reduce the possible biasing effect of ability on the response comparisons.

Key elements of any experiment will be discussed and illustrated in this section. These key elements include the following:

1. Clearly define the *question* to be investigated.
2. Identify the key variables to be used as *treatments.*
3. Identify other important variables that can be *controlled.*
4. Identify important background (lurking) variables that cannot be controlled but should be balanced by *randomization.*
5. Randomly assign treatments to the *experimental units.*
6. Decide on a *method of measurement* that will minimize *measurement bias.*
7. Organize the *data collection* and *data management.*

8. Plan for careful and thorough *data analysis.*
9. Write *conclusions* in the light of the original question.
10. Plan a *follow-up* study to answer the question more completely or to answer the next logical question on the issue at hand.

We will now see how these steps are followed in a real experiment of practical significance.

Does Aspirin Help Prevent Heart Attacks? The Physicians' Health Study

During the 1980s, approximately 22,000 physicians older than age 40 agreed to participate in a long-term health study for which one important question was to determine whether aspirin helps lower the rate of heart attacks (myocardial infarctions). The treatment for this part of the study was aspirin, and the control was a placebo. Physicians were randomly assigned to one treatment or the other as they entered the study to minimize bias caused by uncontrolled factors. The method of assignment was equivalent to tossing a coin and sending the physician to the aspirin arm of the study if a head appeared on the coin.

After the assignment, neither the participating physicians nor the medical personnel who treated them knew who was taking aspirin and who was taking placebo. This is called a *double-blind experiment.* (Why is the double blinding important in a study such as this?) The method of measurement was to observe the physicians carefully for an extended period and to record all heart attacks, as well as other problems, that might occur.

Other than aspirin, many variables could affect the rate of heart attacks for the two groups of physicians. For example, the amount of exercise they get and whether or not they smoke are two prime examples of variables that should be controlled in the study so that the true effect of aspirin can be measured. Table 1 shows how the subjects eventually were divided according to exercise and cigarette smoking. (See Tables 2 through 4 at the end of this section for more details.) Do you think the randomization scheme did a good job in controlling these variables? Would you be concerned about the results for aspirin being unduly influenced by the fact that most of the aspirin takers were also nonsmokers? Would you be concerned about the placebo group possibly having too many physicians who do not exercise?

		Aspirin	Placebo
Exercise Vigorously	Yes	7910	7861
	No	2997	3060
Cigarette Smoking	Never	5431	5448
	Past	4373	4301
	Current	1213	1225

Table 1

The data analysis for this study reports that 139 heart attacks developed among the aspirin users and 239 heart attacks developed in the placebo group. This was said to be a significant result in favor of aspirin as a possible preventive measure for heart attacks. To see why this was so, work through the following steps.

1. Given that there were approximately 11,000 participants in each arm of the study, calculate the proportion of heart attacks among those taking aspirin. Calculate the proportion of heart attacks among those taking placebo.

2. Calculate the standard deviations for each of these proportions and use them to form confidence intervals for the true proportions of heart attacks to be expected among aspirin users and among nonaspirin users in the population from which these sampling units were selected.

3. Looking at the two confidence intervals, can you see why the researchers in this study declared that aspirin had a significant effect in reducing heart attacks? Explain.

   However, heart attacks aren't the only cause for concern. Another is that too much aspirin can cause an increase in strokes. Among the aspirin users on the study, 119 had strokes during the observation period. Within the placebo group, only 98 had strokes. Although the number of strokes is higher than the researchers would have liked, the difference between the two numbers was no cause for alarm. That is, there did not appear to be a significant increase in the number of strokes for the aspirin group. Follow the three previous steps for constructing and observing confidence intervals to see why the researchers were not overly concerned about the difference in numbers of strokes between the two arms of the study.

Much more data relating to this study are provided in Tables 2 through 4. This should lead to other questions of interest regarding the relationships among aspirin use, heart attacks, and other factors. (For example, does age play a role in the effectiveness of aspirin? How about cholesterol level?)

	End Point	Aspirin Group	Placebo Group	Relative Risk	95% Confidence Interval	p Value
Myocardial Infarction	Fatal	10	26	0.34	0.15–0.75	0.007
	Nonfatal	129	213	0.59	0.47–0.74	<0.00001
	Total	139	239	0.56	0.45–0.70	<0.00001
	Person-Years of Observation	54,560.0	54,355.7	—	—	—
Stroke	Fatal	9	6	1.51	0.54–4.28	0.43
	Nonfatal	110	92	1.20	0.91–1.59	0.20
	Total	119	98	1.22	0.93–1.60	0.15

Table 2: Confirmed cardiovascular end points in the aspirin component of the Physicians' Health Study, according to the Treatment Group

Cause*	Aspirin Group	Placebo Group	Relative Risk	95% Confidence Interval	*p* Value
Total cardiovascular deaths[1]	81	83	0.96	0.60–1.54	0.870
Acute myocardial infarction (410)	10	28	0.31	0.14–0.68	0.004
Other ischemic heart disease (411–414)	24	25	0.97	0.60–1.55	0.890
Sudden death (798)	22	12	1.96	0.91–4.22	0.090
Stroke (430, 431, 434, 436)[2]	10	7	1.44	0.54–3.88	0.470
Other cardiovascular (402, 421, 424, 425, 428, 429, 437, 440, 441)	15	11	1.38	0.62–3.05	0.430
Total noncardiovascular deaths	124[3]	133	0.93	0.72–1.20	0.590
Total deaths with confirmed cause	205	216	0.95	0.79–1.15	0.600
Total deaths[4]	217	227	0.96	0.80–1.14	0.640

Table 3: Confirmed deaths, according to Treatment Group

*Numbers are code numbers of the International Classification of Diseases, ninth revision.

[1]All fatal cardiovascular events are included, regardless of previous nonfatal events.

[2]This category includes ischemic (three in the aspirin and three in the placebo group), hemorrhagic (seven aspirin and two placebo), and unknown cause (zero aspirin and two placebo).

[3]This category includes one death due to gastrointestinal hemorrhage.

[4]Additional events that could not be confirmed because records were not available included 23 deaths (12 aspirin and 11 placebo), of which 11 were suspected to be cardiovascular (7 aspirin and 4 placebo) and 12 noncardiovascular (5 aspirin and 7 placebo).

		No. of Myocardial Infarctions/ Total No. (%)			
		Aspirin Group	Placebo Group	Relative Risk	*p* Value of Trend in Relative Risk
Age (yr)	40–49	27/4527 (0.6)	24/4524 (0.5)	1.12	0.02
	50–59	51/3725 (1.4)	87/3725 (2.3)	0.58	
	60–69	39/2045 (1.9)	84/2045 (4.1)	0.46	
	70–84	22/740 (3.0)	44/740 (6.0)	0.49	
Cigarette Smoking	Never	55/5431 (1.0)	96/5488 (1.8)	0.58	0.99
	Past	63/4373 (1.4)	105/4301 (2.4)	0.59	
	Current	21/1213 (1.7)	37/1225 (3.0)	0.57	
Diabetes Mellitus	Yes	11/275 (4.0)	26/258 (10.1)	0.39	0.22
	No	128/10,750 (1.2)	213/10,763 (2.0)	0.60	
Parental History of Myocardial Infarction	Yes	23/1420 (1.6)	39/1432 (2.7)	0.59	0.97
	No	112/9505 (1.2)	192/9481 (2.0)	0.58	
Cholesterol Level (mg per 100 ml)*	<159	2/382 (0.5)	9/406 (2.2)	0.23	0.04
	160–209	12/1587 (0.8)	37/1511 (2.5)	0.29	
	210–259	26/1435 (1.8)	43/1444 (3.0)	0.61	
	≥260	14/582 (2.4)	23/570 (4.0)	0.59	
Diastolic Blood Pressure (mm Hg)	≤69	2/583 (0.3)	9/562 (1.6)	0.21	0.88
	70–79	24/2999 (0.8)	40/3076 (1.3)	0.61	
	80–89	71/5061 (1.4)	128/5083 (2.5)	0.55	
	≥90	26/1037 (2.5)	43/970 (4.4)	0.56	
Systolic Blood Pressure (mm Hg)	<109	1/330 (0.3)	4/296 (1.4)	0.22	0.48
	110–129	40/5072 (0.8)	75/5129 (1.5)	0.52	
	130–149	63/3829 (1.7)	115/3861 (3.0)	0.55	
	≥150	19/454 (4.2)	26/412 (6.3)	0.65	

(continued)

		No. of Myocardial Infarctions/ Total No. (%)			
		Aspirin Group	Placebo Group	Relative Risk	*p* Value of Trend in Relative Risk
Alcohol Use	Daily	26/2718 (1.0)	55/2727 (2.0)	0.45	0.26
	Weekly	70/5419 (1.3)	112/5313 (2.1)	0.61	
	Rarely	40/2802 (1.4)	65/2897 (2.2)	0.63	
Vigorous Exercise at Least Once a Week	Yes	91/7910 (1.2)	140/7861 (1.8)	0.65	0.21
	No	45/2997 (1.5)	92/3060 (3.0)	0.49	
Body Mass Index†	≤23.0126	26/2872 (0.9)	41/2807 (1.5)	0.61	0.90
	23.0127–24.4075	32/2700 (1.2)	46/2627 (1.8)	0.68	
	24.4076–26.3865	32/2713 (1.2)	75/2823 (2.7)	0.44	
	≥26.3866	49/2750 (1.8)	76/2776 (2.7)	0.65	

Table 4: Risk of total myocardial infarction associated with aspirin use, according to level of coronary risk factors

*To convert cholesterol value to millimoles per liter, multiply by 0.02586.

†Body mass index is the weight (in kilograms) divided by the height (in meters) squared.

Modeling

What Is a Model?

Traffic safety depends, in part, on the ability of drivers to anticipate their stopping distance. Stopping distance is related to the speed at which drivers are traveling. In fact, handbooks for drivers have tables of typical stopping distances as a function of speed. How do we know the relationship between the speed of a car and stopping distance? Why does the handbook have just one table rather than a different table for each type of car? How were the data in the table obtained? These are all questions about modeling the relationship between speed and stopping distance for cars.

In this discussion, a *model* refers to a mathematical relationship (often expressed in a formula) among two or more variables. At this introductory level, most examples will involve the study of relationships between only two variables. For example, the area of a circle is related to the radius of the circle by a very specific (and very accurate) formula, which could be called a model for this well-known relationship. Once you know the radius of a circle, you can predict the area from the model; you do not need to measure the area directly. Charts in a pediatrician's office show the relationship between the weight and age of growing children. These relationships also come about through modeling weight as a function of age, but these models are much less accurate than the one involving the radius and area of a circle (why?).

Models, then, are widely used in common situations. We construct models to predict a value not readily observed or to study the possible causal link between variables. The driver (or, at least, the police officer) wants to predict stopping distance by knowing only the speed of the car and wants to see how increased speed will cause the results to change; the driver does not want to conduct his or her own experiments to find out how these two variables are related. The mother and father want to know if their child is experiencing normal weight gain and to predict the weight that the child might attain in 6 more months; they do not want to measure other children, on their own, to establish these facts.

Constructing a Model

The steps in constructing a model are the same as those in the general problem-solving format in the earlier section on quality. First, a clear statement of the problem is essential. Establishing a relationship between speed and stopping distance for cars is too general a question. Driver reaction time plays a key role in making quick stops—is the model concerned with distance from the point at which the driver sees an emergency situation or from the point at which brakes are applied? (What other conditions would have to be considered and qualified?)

Under the conditions set in the problem statement, data are collected and analyzed. Data for establishing a model for stopping distance versus speed come from highly controlled experiments on test tracks, using vehicles of a variety of sizes.

Thus, the model reported is a sort of average for various vehicles, but it is usually stated that the figures are for dry pavement only (why?).

The interpretation of the data often involves fitting a variety of models to the experimental data and choosing the one that seems to provide the best explanation. Very complicated models could be produced to explain how stopping distance relates to speed, with adjustments for road conditions, weight of vehicle, and other factors, but a model that is simple to interpret and use may work almost as well. So, the "best" model is difficult to define and often depends on the intended uses of the model.

Verifying that the selected model actually works is the next step. Once a model for predicting stopping distance is obtained from experimental data, it should be tried in a new experiment to see if the results are reproducible.

Planning the next actions is something to be considered carefully in modeling problems because, in general, all models can be improved or adjusted to specific cases. A model is an approximation to reality and, hence, should never be considered as the final word.

Body Composition

Health and physical fitness are of vital concern to many people in this day and age. We can choose lifestyles that will drastically alter how we feel and how productive we are; one of the key variables in this process is weight. (Why are there so many diet books on the market?) But to a large extent, weight is determined by a person's size and body structure. A more important related variable that *can* be controlled more directly by diet and exercise is percentage of body fat. The problem is that percentage of body fat is somewhat difficult to measure. This leads directly to a modeling problem. The goal is to find a model that estimates percentage of body fat as a function of easily measured variables.

The best way to measure percentage of body fat is by using a laboratory technique called *hydrostatic weighing* (underwater weighing), but this is expensive in terms of time, equipment, and training of technicians. Many anthropometric body measurements (such as height, weight, and skinfold fat measurements) are, however, relatively easy to obtain. So, the refined goal is to build a model to estimate the percentage of body fat from anthropometric measurements.

Background data from a large number of people varying in age and body fat show, under careful exploration, that some of the anthropometric measurements are more highly correlated with body density (which is related to percentage of fat) than others (see Table 1). In particular, the skinfold measurements show consistently high correlation for both men and women. So these measurements will be used as the basis for constructing the model. The data exploration also shows that men and women have quite different body compositions (see Table 2), and this suggests that we may need two models, one for men and one for women. Age turns out to be important in the relationship between skinfold measurements and percentage of body fat, but weight is not. (Can you rationalize this?)

In using linear regression techniques to construct a model, it is found that the sum of three different skinfold measurements works about as well as the sum of all seven. In the spirit of keeping the model simple so that it can be used and will be used correctly, it is decided that the three-measurement model is the "best." However, the three measurements used for men differ from the three measurements used for women.

Variables	Men (n = 402)	Women (n = 283)
Height	−0.03	−0.06
Weight	−0.63	−0.63
Body mass index*	−0.69	−0.70
Skinfolds		
Chest	−0.85	−0.64
Axilla	−0.82	−0.73
Triceps	−0.79	−0.77
Subscapula	−0.77	−0.67
Abdomen	−0.83	−0.75
Suprailium	−0.76	−0.76
Thigh	−0.78	−0.74
Sum of seven	−0.88	−0.83
Circumferences		
Waist	−0.80	−0.71
Gluteal	−0.69	−0.74
Thigh	−0.64	−0.68
Biceps	−0.51	−0.63
Forearm	−0.35	−0.41

Table 1: Linear correlations between body density and anthropometric variables for adults

*Body mass index is the mass (in kilograms) divided by the height (in meters) squared.

	Variables	Men (n = 402) Mean	SD	Range	Women (n = 283) Mean	SD	Range
General Characteristics	Age (yr)	32.8	11.0	18–61	31.8	11.5	18–55
	Height (cm)	179.0	6.4	163–201	168.6	5.8	152–185
	Weight (kg)	78.2	11.7	53–123	57.5	7.4	36–88
	Body mass index (wt/ht^2)	24.4	3.2	17–37	20.2	2.2	14–31
Laboratory Determined	Body density (gm/ml)	1.058	0.018	1.016–1.100	1.044	0.016	1.022–1.091
	Percent fat (%)	17.9	8.0	1–37	24.4	7.2	8–44
	Lean weight (kg)	63.5	7.3	47–100	43.1	4.2	30–54
	Fat weight (kg)	14.6	7.9	1–42	14.3	5.7	2–35
Skinfolds (mm)	Chest	15.2	8.0	3–41	12.6	4.8	3–26
	Axilla	17.3	8.7	4–39	13.0	6.1	3–33
	Triceps	14.2	6.1	3–31	18.2	5.9	5–41
	Subscapula	16.0	7.0	5–45	14.2	6.4	5–41
	Abdomen	25.1	10.8	5–56	24.2	9.6	4–36
	Suprailium	16.2	8.9	3–53	14.0	7.1	3–40
	Thigh	18.9	7.7	4–48	29.5	8.0	7–53
Sum of Skinfolds (mm)	All seven	122.9	52.0	31–272	125.6	42.0	35–266
	Chest, abdomen, thigh	59.2	24.5	10–118			
	Triceps, chest, subscapula	45.3	19.6	11–105			
	Triceps, suprailium, thigh				61.6	19.0	16–126
	Triceps, suprailium, abdomen				56.3	21.0	13–131

Table 2: Descriptive statistics and statistical differences for men and women

Age to Last Year

Sum of Skinfolds (mm)	22 and Younger	23–27	28–32	33–37	38–42	43–47	48–52	53–57	Older Than 57
8–10	1.3	1.8	2.3	2.9	3.4	3.9	4.5	5.0	5.5
11–13	2.2	2.8	3.3	3.9	4.4	4.9	5.5	6.0	6.5
14–16	3.2	3.8	4.3	4.8	5.4	5.9	6.4	7.0	7.5
17–19	4.2	4.7	5.3	5.8	6.3	6.9	7.4	8.0	8.5
20–22	5.1	5.7	6.2	6.8	7.3	7.9	8.4	8.9	9.5
23–25	6.1	6.6	7.2	7.7	8.3	8.8	9.4	9.9	10.5
26–28	7.0	7.6	8.1	8.7	9.2	9.8	10.3	10.9	11.4
29–31	8.0	8.5	9.1	9.6	10.2	10.7	11.3	11.8	12.4
32–34	8.9	9.4	10.0	10.5	11.1	11.6	12.2	12.8	13.3
35–37	9.8	10.4	10.9	11.5	12.0	12.6	13.1	13.7	14.3
38–40	10.7	11.3	11.8	12.4	12.9	13.5	14.1	14.6	15.2
41–43	11.6	12.2	12.7	13.3	13.8	14.4	15.0	15.5	16.1
44–46	12.5	13.1	13.6	14.2	14.7	15.3	15.9	16.4	17.0
47–49	13.4	13.9	14.5	15.1	15.6	16.2	16.8	17.3	17.9
50–52	14.3	14.8	15.4	15.9	16.5	17.1	17.6	18.2	18.8
53–55	15.1	15.7	16.2	16.8	17.4	17.9	18.5	19.1	19.7
56–58	16.0	16.5	17.1	17.7	18.2	18.8	19.4	20.0	20.5
59–61	16.9	17.4	17.9	18.5	19.1	19.7	20.2	20.8	21.4
62–64	17.6	18.2	18.8	19.4	19.9	20.5	21.1	21.7	22.2
65–67	18.5	19.0	19.6	20.2	20.8	21.3	21.9	22.5	23.1
68–70	19.3	19.9	20.4	21.0	21.6	22.2	22.7	23.3	23.9
71–73	20.1	20.7	21.2	21.8	22.4	23.0	23.6	24.1	24.7
74–76	20.9	21.5	22.0	22.6	23.2	23.8	24.4	25.0	25.5
77–79	21.7	22.2	22.8	23.4	24.0	24.6	25.2	25.8	26.3
80–82	22.4	23.0	23.6	24.2	24.8	25.4	25.9	26.5	27.1
83–85	23.2	23.8	24.4	25.0	25.5	26.1	26.7	27.3	27.9
86–88	24.0	24.5	25.1	25.7	26.3	26.9	27.5	28.1	28.7
89–91	24.7	25.3	25.9	26.5	27.1	27.6	28.2	28.8	29.4
92–94	25.4	26.0	26.6	27.2	27.8	28.4	29.0	29.6	30.2
95–97	26.1	26.7	27.3	27.9	28.5	29.1	29.7	30.3	30.9
98–100	26.9	27.4	28.0	28.6	29.2	29.8	30.4	31.0	31.6
101–103	27.5	28.1	28.7	29.3	29.9	30.5	31.1	31.7	32.3
104–106	28.2	28.8	29.4	30.0	30.6	31.2	31.8	32.4	33.0
107–109	28.9	29.5	30.1	30.7	31.3	31.9	32.5	33.1	33.7
110–112	29.6	30.2	30.8	31.4	32.0	32.6	33.2	33.8	34.4
113–115	30.2	30.8	31.4	32.0	32.6	33.2	33.8	34.5	35.1
116–118	30.9	31.5	32.1	32.7	33.3	33.9	34.5	35.1	35.7
119–121	31.5	32.1	32.7	33.3	33.9	34.5	35.1	35.7	36.4
122–124	32.1	32.7	33.3	33.9	34.5	35.1	35.8	36.4	37.0
125–127	32.7	33.3	33.9	34.5	35.1	35.8	36.4	37.0	37.6

Table 3: Percent fat estimate for men: sum of chest, abdomen, and thigh skinfolds

Age to Last Year

Sum of Skinfolds (mm)	22 and Younger	23–27	28–32	33–37	38–42	43–47	48–52	53–57	Older Than 57
23–25	9.7	9.9	10.2	10.4	10.7	10.9	11.2	11.4	11.7
26–28	11.0	11.2	11.5	11.7	12.0	12.3	12.5	12.7	13.0
29–31	12.3	12.5	12.8	13.0	13.3	13.5	13.8	14.0	14.3
32–34	13.6	13.8	14.0	14.3	14.5	14.8	15.0	15.3	15.5
35–37	14.8	15.0	15.3	15.5	15.8	16.0	16.3	16.5	16.8
38–40	16.0	16.3	16.5	16.7	17.0	17.2	17.5	17.7	18.0
41–43	17.2	17.4	17.7	17.9	18.2	18.4	18.7	18.9	19.2
44–46	18.3	18.6	18.8	19.1	19.3	19.6	19.8	20.1	20.3
47–49	19.5	19.7	20.0	20.2	20.5	20.7	21.0	21.2	21.5
50–52	20.6	20.8	21.1	21.3	21.6	21.8	22.1	22.3	22.6
53–55	21.7	21.9	22.1	22.4	22.6	22.9	23.1	23.4	23.6
56–58	22.7	23.0	23.2	23.4	23.7	23.9	24.2	24.4	24.7
59–61	23.7	24.0	24.2	24.5	24.7	25.0	25.2	25.5	25.7
62–64	24.7	25.0	25.2	25.5	25.7	26.0	26.7	26.4	26.7
65–67	25.7	25.9	26.2	26.4	26.7	26.9	27.2	27.4	27.7
68–70	26.6	26.9	27.1	27.4	27.6	27.9	28.1	28.4	28.6
71–73	27.5	27.8	28.0	28.3	28.5	28.8	29.0	29.3	29.5
74–76	28.4	28.7	28.9	29.2	29.4	29.7	29.9	30.2	30.4
77–79	29.3	29.5	29.8	30.0	30.3	30.5	30.8	31.0	31.3
80–82	30.1	30.4	30.6	30.9	31.1	31.4	31.6	31.9	32.1
83–85	30.9	31.2	31.4	31.7	31.9	32.2	32.4	32.7	32.9
86–88	31.7	32.0	32.2	32.5	32.7	32.9	33.2	33.4	33.7
89–91	32.5	32.7	33.0	33.2	33.5	33.7	33.9	34.2	34.4
92–94	33.2	33.4	33.7	33.9	34.2	34.4	34.7	34.9	35.2
95–97	33.9	34.1	34.4	34.6	34.9	35.1	35.4	35.6	35.9
98–100	34.6	34.8	35.1	35.3	35.5	35.8	36.0	36.3	36.5
101–103	35.3	35.4	35.7	35.9	36.2	36.4	36.7	36.9	37.2
104–106	35.8	36.1	36.3	36.6	36.8	37.1	37.3	37.5	37.8
107–109	36.4	36.7	36.9	37.1	37.4	37.6	37.9	38.1	38.4
110–112	37.0	37.2	37.5	37.7	38.0	38.2	38.5	38.7	38.9
113–115	37.5	37.8	38.0	38.2	38.5	38.7	39.0	39.2	39.5
116–118	38.0	38.3	38.5	38.8	39.0	39.3	39.5	39.7	40.0
119–121	38.5	38.7	39.0	39.2	39.5	39.7	40.0	40.2	40.5
122–124	39.0	39.2	39.4	39.7	39.9	40.2	40.4	40.7	40.9
125–127	39.4	39.6	39.9	40.1	40.4	40.6	40.9	41.1	41.4
128–130	39.8	40.0	40.3	40.5	40.8	41.0	41.3	41.5	41.8

Table 4: Percent fat estimate for women: sum of triceps, suprailium, and thigh skinfolds

Data analysis using regression techniques produced the estimates of percentage of body fat as a function of the sum of three skinfold measurements seen in Tables 3 and 4. These estimates came from models that made use of the hydrostatically produced fat percentages as "truth" and attempted to reproduce these values as functions of the skinfold data. Although we do not have the original data, we can recapture the flavor of these models by fitting simple regression models to the data found in Table 3 or 4. From Table 3 (using the midpoint of ranges as the point observation), the relationship between percentage of body fat (Y) and the sum of the three skinfolds (X) for the 22 and younger age category is

$$Y = 0.013 + 0.270X$$

and the model for the 53 to 57 age category is

$$Y = 3.510 + 0.282X$$

For the women (Table 4), the corresponding models for those 22 and younger are

$$Y = 5.470 + 0.287X$$

and for women 53 to 57

$$Y = 7.210 + 0.287X$$

(Interpret these fitted models in terms of the variables under discussion. How are the models the same, and how do they differ? What is the practical significance of the similarities and differences?)

In extending the preceding analysis, one should study the residual plots for these fitted models. These plots might suggest that the simple linear models are not quite accurate, although they may be adequate over certain ranges of X.

How will the estimates of percentage of body fat be used? One possibility is to aid in setting weight-reduction goals. If we know our current weight and percentage of body fat and have a desired body fat percentage in mind, we can calculate our desired weight with this formula:

$$\text{Desired weight} = \frac{\text{Weight} - \left[\text{Weight}\left(\dfrac{\%\text{ fat}}{100}\right)\right]}{1 - \left[\dfrac{\%\text{ fat desired}}{100}\right]}$$

Note that current weight *is* important in estimating a desired weight goal. (Can you explain why the formula works? Are any assumptions being made here?)

What is a reasonable percentage of body fat to set as a goal? Some recommend between 10% and 22% for men and between 20% and 32% for women. This, however, is not good enough for athletes, who should average around 12% if male and 18% if female.

REFERENCES

Matching Descriptions to Scatter Plots

1. F. Anscombe (1973), "Graphs in statistical analysis," *American Statistician,* 27:17–21. (This is the source of the data sets in "Extensions.")
2. The marriage rate and divorce rate data come from J. M. Landwehr and A. E. Watkins (1995), *Exploring Data,* rev. ed., Palo Alto, CA: Dale Seymour Publications. Their sources were the United Nations publications *Demographic Yearbook 1990* and *Monthly Bulletin of Statistics,* June 1991.
3. The Anscombe data sets are discussed in S. Weisberg (1985), *Applied Linear Regression,* 2nd ed., New York: Wiley—and in many other texts.

The Regression Effect

1. D. Freedman et al. (1991), *Statistics,* 2nd ed., New York: Norton, pp. 159–165.
2. D. Kahneman, P. Slovic, and A. Tversky, eds. (1982), *Judgment under Uncertainty: Heuristics and Biases,* Cambridge, MA: Cambridge University Press. Psychologists have found that people don't recognize the regression effect and that if they do notice it, they tend to feel that it needs explanation.
3. J. R. Levin (1993), "An improved modification of a regression-toward-the-mean demonstration," *American Statistician,* 47(February):24–26. This article describes how to use two decks of cards to illustrate the regression effect. The basic idea is this: Divide one deck into a lower part and an upper part—all 1 (ace) through 6 and all 8 through 13 (king). Discard the 7's. Half of the class depict low scores and draw their "true" score from the top half of the deck. Each member of the class then draws his or her "error" from the second deck and adds it to his or her "true" score. On the retest, all students retain their "true" score and draw another "error" from the second deck.
4. R. W. Mee and T. Chiu Chua (1991), "Regression toward the mean and the paired sample *t* test," *American Statistician,* 45(February):39–42. This article shows how one should properly conduct a paired-sample comparison of means in a test-retest situation where only a subset of the population retakes the exam. For example, suppose all who fail an exam are given the opportunity to retake a similar exam. If the usual paired-sample *t* test is conducted from data in this setting, the "regression effect" may lead to the incorrect conclusion that some intervention (such as remedial tutoring) has been effective in raising the scores when this is not necessarily true.
5. A. E. Watkins (1986), "The regression effect; or, I always thought that the rich get richer . . . ," *The Mathematics Teacher,* 79:644–647.

Leonardo's Model Bodies

1. M. Kemp, ed. (1989), *Leonardo on Painting,* New Haven, CT: Yale University Press.

Models, Models, Models

1. Atmospheric CO_2 concentrations—Mauna Loa Observatory, Hawaii, 1958–1986, C. D. Keeling, Scripps Institution of Oceanography. CDIAC NDP-001/R1 (rev. 1986).
2. *Basic Petroleum Data Book.* Petroleum Industry Statistics, Volume XIV, Number 1, January 1994, American Petroleum Institute, Washington, D.C.

Predictable Pairs

1. "The final report on the aspirin component of the ongoing Physicians' Health Study," *N. Engl. J. Med.* (1989), 231(3):129–135.
2. K. Allen and A. Moss (1993), "Teenage tobacco use," *Advance Data,* no. 224 (Feb. 1), National Center for Health Statistics.

Ratings and Ranks

1. R. W. Johnson (1993), "How does the NFL rate the passing ability of quarterbacks?" *Coll. Math. J.,* 24:451–453.
2. D. Savageau and R. Boyer (1993), *Places Rated Almanac,* New York: Prentice Hall Travel.

Gummy Bears in Space

1. S. M. Kosslyn (1980), *Image and Mind,* Cambridge, MA: Harvard University Press, Chapter 9.

Funnel Swirling

1. B. Gunter (1993), "Through a funnel slowly with ball bearings and insight to teach experimental design," *American Statistician,* 47:265–269.

Jumping Frogs

1. R. L. Scheaffer (1989), *Planning and Analyzing Experiments,* unpublished manuscript.

How to Ask Questions: Designing a Survey

1. American Statistical Association (1994), *What Is a Survey?* In J. D. Cryer and R. B. Miller, *Statistics for Business: Data Analysis and Modeling,* 2nd ed., Belmont, CA: Duxbury, pp. 388–399.
2. D. S. Moore (1991), *Statistics: Concepts and Controversies,* 3rd ed., San Francisco: Freeman.

Most college libraries carry books on the design of surveys. Three of the many good ones are these:

3. C. A. Moser and G. Kalton (1972), *Survey Methods in Social Investigation,* 2nd ed., New York: Basic Books.
4. A. N. Oppenheim (1966), *Questionnaire Design and Attitude Measurement,* New York: Basic Books.
5. H. Schuman and S. Presser (1981), *Questions and Answers in Attitude Surveys,* New York: Academic Press.

What Is Random Behavior?

1. J. E. Cohen (1973), "Turning the tables," *Statistics by Example: Exploring Data,* Reading, MA: Addison-Wesley, pp. 87–90.

The Law of Averages

1. W. Feller (1968), *An Introduction to Probability Theory and Its Applications,* vol. I, 3rd ed., New York: Wiley.
2. D. Freedman et al. (1991), *Statistics,* 2nd ed., New York: W.W. Norton, p. 249.
3. M. Hollander and F. Proschan (1984), *The Statistical Exorcist: Dispelling Statistics Anxiety,* New York: Marcel Dekker, pp. 203–204.
4. D. Moore (1985), *Statistics: Concepts and Controversies,* 2nd ed., New York: Freeman, pp. 266–267.
5. A. E. Watkins (1995), "The law of averages," *Chance,* Spring.

Streaky Behavior

Students can read and report on some of these articles:

1. "Slumps," *Money* (February 1985), p. 12.
2. C. Albright (1992), "Streaks & Slumps," *OR/MS Today* (April):94–95. The author ends, "For now, we simply conclude by stating that if there are perennially streaky hitters, the present analysis was not able to find them."
3. S. C. Albright (1993), "A statistical analysis of hitting streaks in baseball (with discussion)," *American Statistician,* 88:1175–1183.
4. R. Hooke (1989), "Basketball, baseball and the null hypothesis," *Chance,* 2(4):35–37.
5. P. D. Larkey et al. (1989). "It's okay to believe in the 'hot hand,'" *Chance,* 2(4):22–30.
6. M. F. Schilling (1990), "The longest run of heads," *Coll. Math. J.,* 21:196–207.
7. M. F. Schilling (1994), "Long-run predictions," *Math Horizons,* Spring, pp. 10–12. Schilling's article gives the formulas for the probability distributions of the longest run of successes (and failures) for different numbers of trials and probability of success.
8. A. Tversky and T. Gilovich (1989), "The 'hot hand': statistical reality or cognitive illusion?" *Chance,* 2(4):31–34.
9. A. Tversky and T. Gilovich (1989), "The cold facts about the 'hot hand' in basketball," *Chance,* 2(1):16–21.

Counting Successes

1. M. Gnanadesikan, R. L. Scheaffer, and J. Swift (1987), *The Art and Techniques of Simulation,* Palo Alto, CA: Dale Seymour Publications.

Waiting for Sammy Sosa

1. R. J. Larsen and M. L. Marx (1986), *An Introduction to Mathematical Statistics and Its Applications,* 2nd ed., Englewood Cliffs, NJ: Prentice Hall, pp. 218–221.
2. F. Mosteller, R. E. K. Rourke, and G. B. Thomas, Jr. (1970), *Probability with Statistical Applications,* 2nd ed., Reading, MA: Addison-Wesley, pp. 176, 189, 219.

Spinning Pennies

1. G. Giles (1986), "The Stirling recording sheet for experiments in probability." In P. Holmes, ed., *The Best of Teaching Statistics,* Sheffield, England: Teaching Statistics Trust, pp. 8–14.

How Many Tanks?

1. D. C. Flaspohler and A. L. Dinkheller (1999), "German tanks: a problem in estimation," *The Mathematics Teacher,* November, pp. 724–728.
2. R. W. Johnson (1994), "Estimating the size of a population," *Teaching Statistics,* 16(2):50–52.
3. J. M. Landwehr, J. Swift, and A. E. Watkins (1987), *Exploring Surveys and Information from Samples,* Palo Alto, CA: Dale Seymour Publications, pp. 75–83.
4. R. Ruggles and H. Brodie (1947), "An empirical approach to economic intelligence in World War II," *J. Amer. Stat. Assoc.,* 42:72–91.

What Is a Confidence Interval Anyway?

1. J. M. Landwehr, J. Swift, and A. E. Watkins (1987), *Exploring Surveys and Information from Samples,* Palo Alto, CA: Dale Seymour Publications.

Confidence Intervals for the Proportion of Even Digits

1. American Statistical Association. "What Is a Survey?" Reprinted in J. D. Cryer and R. B. Miller (1994), *Statistics for Business: Data Analysis and Modeling,* 2nd ed., Belmont, CA: Duxbury, pp. 388–399.
2. D. Freedman et al. (1991), *Statistics,* 2nd ed., New York: W.W. Norton.
3. J. M. Landwehr, J. Swift, and A. E. Watkins (1986), *Exploring Surveys: Information from Samples,* Menlo Park, CA: Dale Seymour Publications.
4. D. S. Moore (1985), *Statistics: Concepts and Controversies,* 2nd ed., New York: Freeman.

Capture/Recapture

1. B. Bailar (1988), "Who counts in America?" *Chance,* 1:9, 17.
2. D. G. Chapman (1989), "The plight of the whales." In J. M. Tanur et al. (eds.), *Statistics: A Guide to the Unknown,* 3rd ed., Belmont, CA: Wadsworth, pp. 60–67.
3. E. P. Erickson and J. B. Kadane (1985), "Estimating the population in a census year: 1980 and beyond" (with discussion), *J. Amer. Stat. Assoc.,* 80:98–131.
4. S. Fienberg (1992), "An adjusted census for 1990? The trial," *Chance,* 5:28–38.
5. H. Hogan (1992), "The 1990 post-enumerative survey: an overview," *American Statistician,* 46:261–269.
6. W. E. Ricker (1975), *Computation and Interpretation of Biological Statistics of Fish Populations,* Bulletin of the Fisheries Board of Canada 191, Ottawa, Canada, pp. 83–86.

How to Ask Sensitive Questions

1. S. Boxer (1987). "AIDS and epidemiology. Women and drugs," *Discover,* 8(7):12.
2. S. L. Warner (1966), "Randomized response: a survey technique for eliminating evasive answer bias," *J. Amer. Stat. Assoc.,* 60:63–69.

The Bootstrap

1. B. Efron (1982), *The Jackknife, the Bootstrap, and Other Resampling Plans,* Philadelphia: Society for Industrial and Applied Mathematics.
2. B. Efron and P. Diaconis (1983), "Computer-intensive methods in statistics," *Scientific American,* 248(5):116–130.
3. B. Efron and J. Tibshirani (1993), *An Introduction to the Bootstrap,* New York: Chapman & Hall.
4. P. Hall (1992), *The Bootstrap and Edgeworth Expansion,* New York: Springer-Verlag.

Statistical Evidence of Discrimination

1. P. Barbella, L. J. Denby, and J. M. Landwehr (1990), "Beyond exploratory data analysis: the randomization test," *The Mathematics Teacher,* 83(February):144–149.
2. F. Mosteller and Robert E. K. Rourke (1973), *Sturdy Statistics: Nonparametric and Order Statistics,* Reading, MA: Addison-Wesley, pp. 12–23.
3. A. Schachter (1959), *The Psychology of Affiliation,* Stanford, CA: Stanford University Press.
4. S. L. Zabell (1989), "Statistical proof of employment discrimination." In Judith M. Tanur et al. (eds.), *Statistics: A Guide to the Unknown,* 3rd ed., Pacific Grove, CA: Wadsworth, pp. 79–86.
5. E. S. Edgington, "Randomization tests," UMAP Module 487, COMAP, Lexington, MA 02173 (phone 617-862-7878).

6. E. S. Edgington (1987), *Randomization Tests,* 2nd ed., New York: Marcel Dekker.
7. R. J. Larsen and D. Fox Stroup (1976), *Statistics in the Real World: A Book of Examples,* New York: Macmillan, pp. 205–207.
8. P. Meier, J. Sacks, and S. L. Zabell (1984), "What happened in Hazelwood: statistics, employment discrimination, and the 80% rule," *American Bar Foundation Research Journal,* (Winter):139–186.
9. J. A. Rice (1988), *Mathematical Statistics and Data Analysis,* Pacific Grove, CA: Wadsworth, pp. 434–436, 452.
10. S. L. Zabell (1989), "Statistical proof of employment discrimination." In Judith M. Tanur et al. (eds.), *Statistics: A Guide to the Unknown,* 3rd ed., Pacific Grove, CA: Wadsworth, pp. 79–86.

Application: Body Composition

1. A. S. Jackson and M. L. Pollock, (1985), "Practical assessment of body composition," *The Physician and Sports Medicine,* 13(5) May.

INDEX

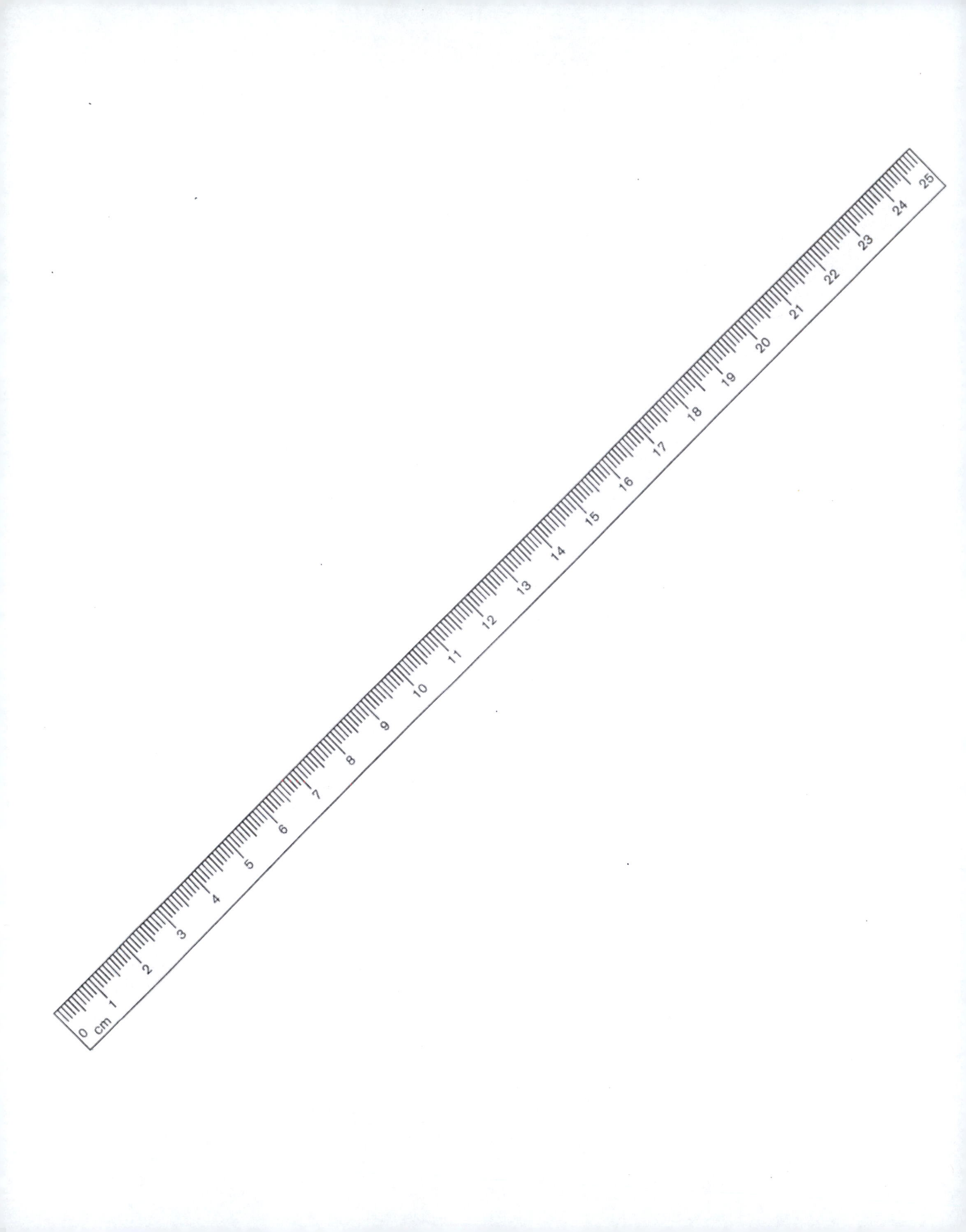
0 cm
1
2
3
4
5
6
7
8
9
10
11
12
13
14
15
16
17
18
19
20
21
22
23
24
25

CPSIA information can be obtained at www.ICGtesting.com
Printed in the USA
BVOW081317290112

281585BV00006B/2/P